WAKE TECHNICAL COMMUNITY COLLEGE LIBRARY
9101 FAYETTEVILLE ROAD
RALEIGH, NORTH CAROLINA 27603

TECHNOLOGY, TRANSPORT, AND TRAVEL IN AMERICAN HISTORY

**AMERICAN HISTORICAL ASSOCIATION–
SOCIETY FOR THE HISTORY OF TECHNOLOGY**

HISTORICAL PERSPECTIVES ON TECHNOLOGY, SOCIETY, AND CULTURE

A SERIES EDITED BY PAMELA O. LONG AND ROBERT C. POST

OTHER TITLES PUBLISHED:

TECHNOLOGY AND SOCIETY IN MING CHINA (1368–1644)
Francesca Bray

TECHNOLOGY, SOCIETY, AND CULTURE IN LATE MEDIEVAL AND RENAISSANCE EUROPE, 1300–1600
Pamela O. Long

THE MILITARY INDUSTRIAL COMPLEX
Alex Roland

TECHNOLOGY TRANSFER AND EAST ASIAN ECONOMIC TRANSFORMATION
Rudi Volti

TITLES FORTHCOMING:

MILITARY TECHNOLOGY, MILITARY INSTITUTIONS, AND WORLD HISTORY
Barton C. Hacker

ADAPTING TO A CHANGING WORLD: MEDIEVAL TECHNOLOGY, 300–1300
John Muendel, Bert S. Hall, and Pamela O. Long

NATURE AND TECHNOLOGY IN HISTORY
Sara B. Prichard and Jim Williams

Technology, Transport, and Travel in American History

by Robert C. Post

A Publication of the Society for the History of Technology
and the American Historical Association

Robert C. Post is curator emeritus at the National Museum of American History. A former president of the Society for the History of Technology, he was editor-in-chief of *Technology and Culture* from 1981 through 1995, and is now its book review editor. His publications include *High Performance: The Culture and Technology of Drag Racing, 1950–2000* (Baltimore: Johns Hopkins University Press, 2001).

Cover Illustration: A freight train is seen on the original transcontinental railway at Grand Island, Nebraska, with one of the few steam locomotives still in service in the United States in the fall of 1958. The handcart is reminiscent of conveyances used by Mormons a century before to transport belongings across the plains to their Zion at Great Salt Lake. Author's photo.

Notes on usage:
"America" and "American" provide a convenient shorthand, but the terms are, of course, flawed. To borrow an observation from Todd Shallat, readers should keep in mind the "prideful and provincial way the first citizens of the United States of America once used those words, as if people south or north of the border were invisible or inconsequential" (*Structures in the Stream: Water, Science, and the U.S. Army Corps of Engineers* [Austin: University of Texas Press, 1994], x).

Historically, conveyances of several kinds, but most notably oceangoing ships, were coded as female, even if named for a man, and that is the usage here. For persuasive explanations, including domestic (a vessel to be "commanded") and sexual (something to "ride" through a heavy sea), see Margaret Creighton, "Sailing between a Rock and a Hard Place: Navigating Manhood in the 1800s," in Benjamin W. Labaree et al., *America and the Sea: A Maritime History* (Mystic, Conn.: Mystic Seaport, 1998), 293–296.

Layout: Christian A. Hale

This booklet series is published with financial assistance from the Dibner Fund, Inc.

© 2003 American Historical Association

All rights reserved. No part of this book may be reproduced in any form without permission in writing from the publisher, except by a reviewer who wishes to quote brief passages in connection with a review written for inclusion in a magazine or newspaper.

Published in 2003 by the American Historical Association. As publisher, the American Historical Association does not adopt official views on any field of history and does not necessarily agree or disagree with the views expressed in this book.

Library of Congress Cataloging-in-Publication data:
Post, Robert C.
 Technology, Transport, and Travel in American history / by Robert C. Post.
 p. cm—(Historical perspectives on technology, society, and culture)
 Includes bibliographical references.
 ISBN 0-87229-131-6 (pamphlet)
 1. Transportation—Social aspects–United States—History. 2. Technology—Social aspects—United States—History. I. American Historical Association. II. Title. III. Series.

HE203 .P73 2003
388'.0973--dc21   2002154603

TABLE OF CONTENTS

SERIES INTRODUCTION vii

INTRODUCTION 1

1. THE WESTERN OCEAN 5

2. A NEW NATION 9

3. SEAWAYS 17

4. CANALS 25

5. AN INLAND SEACOAST? 31

6. RIVERWAYS 33

7. RAILWAYS 43

8. HIGHWAYS 53

9. AIRWAYS 61

CONCLUSION 69

NOTES 75

SUGGESTED READINGS 95

SERIES INTRODUCTION

Technology reflects and shapes human history. From hunting and gathering cultures and the establishment of neolithic villages, farming, and food storage techniques to the development of metallurgy, ceramics, and weaving; firearms, printing, and mechanized power; and automation, electronics, and computers, history and technology have been integral with one another. The role and function of specific technologies—flint tools in the paleolithic and pottery in the neolithic, the stirrup in the Middle Ages, gunpowder and the mechanical clock in the thirteenth century, printing presses in the fifteenth and sixteenth, the steam engine in the eighteenth, factories in the nineteenth, and the automobile and nuclear power in the twentieth—are all subjects of an expansive scholarly literature. Throughout this literature are animated controversies concerning the choices made among competing techniques for attaining the same end—whether automobiles would be powered by steam, electricity, or internal combustion, for example, or whether computers would be analog or digital.

Yet for all its importance, technology and its mutual interactions with society and culture are rarely addressed in high school, or college, or even in university history courses. When scholars unfamiliar with its rich historiography do consider technology, they typically treat it as inert or determinate, lending their authority to the fallacy that it advances according to its own internal logic. Specialists in the history of technology now recognize the importance of "social construction": technologies succeed or fail (or emerge at all) partly because of the political strategies employed by individual, group, and organizational "actors" who have conflicting or complementary interests in particular outcomes. Many of us believe that success and failure is contingent on inescapable physical realities as well as ambient sociocultural factors. But there is no doubt that technological designs are shaped by such factors; nor, indeed, that the shaping of technology is *integral* to the shaping of society and culture.

This joint venture of the American Historical Association and the Society for the History of Technology draws on the analytical insights of scholars who address technology in social and cultural context, whether their discipline be history or another field in the humanities or social sciences. Authors of these booklets may be concerned with the effects of particular technologies on particular constituencies; with the relationship of technology to labor, economics, and the organization of production; with the role technology plays in differentiating social status and the construction of gender; or with interpretive flexibility—namely, the perception that determinations about whether a technology "works" are contingent on the expectations, needs, and ideology of those who interact with it. Following from this is the understanding that technology is not intrinsically useful or even rational; capitalist ideology in particular has served to mask powerful nonutilitarian motives for technological novelty, among them kinesthetic pleasure, a sense of play, curiosity, and the exercise of ingenuity for its own sake, a phenomenon known as technological enthusiasm. As evidence of this, many modern inventions—from the telephone to the automobile to new materials such as celluloid and aluminum—met only marginal needs at the outset. Needs with any substantial economic significance often had to be *contrived*, thereby making invention the mother of necessity.

There are many definitions of technology. Often they are ahistorical, particularly those that define technology in terms of applying science to industrial and commercial objectives. Sometimes technology is defined as the way that "things are done or made." While this is not a historian's definition per se, it becomes that whenever one asks how things were done or made in a particular way in a particular context and then analyzes the implications of taking one path rather than another. Lynn White Jr., a historian who served as president of both the Society for the History of Technology and the American Historical Association, called this "the jungle of meaning." While the notion that technology marches of its own predetermined accord still has a strong hold on popular sensibilities, specialists in the interaction of technology and culture now understand that it cannot do anything of the sort. Technology is not autonomous; rather it is impelled by choices made in the context of circumstances in ambient realms, very often in the context of disputes over political power. Once chosen, however, technologies themselves can exert a powerful influence on future choices. One only needs to consider the Strategic Defense Initiative, "Star Wars," which keeps getting funded not because it is actually feasible but because it provides partisans with effective political rhetoric.

To some extent, definitions of technology vary from one discipline to another. We believe that defining it as "the sum of the methods by which a social group provides themselves with the material objects of their civilization" is sufficiently concrete without being too confining and without being misleading. It is important to specify the word *material*, for there are of course "techniques" having to do with everything from poetics to sex to bureaucratic administration. Some might go further and specify that "material" be taken to mean three-dimensional "things," and this seems satisfactory as long as one bears in mind that even an abstraction such as a computer program, or an idea for the design of a machine, or an ideology such as technocracy or scientific management is contingent for expression upon tangible artifacts.

"Technology" is a modern word, dating to the early nineteenth century. Its first well-known usage was in an 1831 treatise by Jacob Bigelow entitled *Elements of Technology*. For some time after that—and maybe even today—it was not a term known to every culture. "Mechanical arts," used in medieval and early modern Europe, is not entirely synonymous, since this term included painting and sculpture as well as machinery, mills, and the like. Technology encompasses various actors' categories in diverse historical cultures, and that is part of the reason why contemporary scholars define it variously. We believe that the complexity of definition, conceptual categories, and methodologies is instrumental in making the history of technology such a fascinating and fruitful area of inquiry. In these booklets, each author may be grounding his or her inquiry on somewhat different assumptions about the nature of the subject matter.

"Every generation writes its own history," said Carl Becker. In commissioning and editing these essays, we have sought to have each one convey a broadly informed synthesis of the best scholarship, to outline the salient historiographical issues, and to highlight interpretive stances that seem persuasive to our own generation. We believe that historians of technology are poised to integrate their inquiries with mainstream scholarship, and we trust that these booklets provide ample confirmation of this belief.

Pamela O. Long
Robert C. Post
Series Editors

INTRODUCTION

This booklet is concerned with icons of American history: the storied windjammer, taut canvas everywhere, on a long reach somewhere beyond Cape Horn; the western riverboat, all fancy woodwork and flying sparks, churning down the Mississippi; the Iron Horse invading the Garden of agrarian mythology. But it is concerned with iconography only in part. While addressing travel and transport in the United States, it also aims to show how ambient currents impel technological change in changing historical circumstances—how the deployment of any technology or technological system is a product of cultural, economic, and, perhaps most important, political influences. To simplify a rich debate, we may call this social construction.

Not that technological change cannot have political *consequences* as well. Indeed, it was in response to dangers associated with a new technology in the early 19th century, the high-pressure steam engine, that Congress first enacted legislation empowering the government to coerce private enterprise and exact punishment for misbehavior, a precedent echoed in controversies involving the Interstate Commerce Commission and, more recently, the Occupational Safety and Health Administration (OSHA), the Environmental Protection Agency (EPA), and other federal agencies. But people often apprehend technology as an active and indeed *autonomous* agent of change. Time and time again we get news about a force whose forward motion, usually called "progress," is irresistible—as is its transformation of the world in which we live. Perhaps mindful of the role of personal computers in our daily affairs, many of us accept this idea without a second thought.

Historians are of two minds. Some take technology to be an abstraction whose "agency" remains dubious. Others detect a tendency, notably in the United States, to invest technology with a momentum that can send it "out of control." While this is arguable, the narrative that follows holds to a premise that technological change is ultimately contingent on the outcome of contests over political power. It assumes the validity of Lynn White Jr.'s remark that "a new device merely opens a door; it does not compel one to enter." Hence, it was not *technology* that launched transoceanic voyages of discovery in the 15th century; a door merely opened. Nor did technology impel railways across deserts in the 19th century, or endow humans with wings in the 20th, or cause any other "transportation revolution." Nor did technology construct the social reality of novel conveyances, windjammers, steamboats, or biplanes, trolley cars or jetliners. Rather, their reality was a product of cultural impulses, and, very often, the result of decisions made by people who had their hands on the levers of political power and could decide where financial resources, whether private or public, would be concentrated—that is, decide which doorways to enter.[1]

Which is not to deny the reality of physical constraints that will trump political power in the end, nor to imply that people with political power always behave sensibly. Besides considering the contextual impulses affecting technological change, a second aim of this booklet is to show that technological change is not invariably linked to a sound calculus of economic gain, or to public necessity, or even to motives that seem altogether rational. Human sentiments such as faith in progress and enthusiasm for novelty have often influenced the history of transportation, and various episodes are highlighted here. For one, there was the "canal mania" of the 1830s, which drove several states to the brink of ruin. Then, there were the steamships and clippers that sprinted across the Atlantic and Pacific in the 1840s and 1850s, marvels of ingenuity but profitable only under such a narrow range of external circumstances that they were bound to run aground. After the Civil War, thousands of miles of railway were built "from somewhere to nowhere" with only the vaguest hope that there would ever be enough traffic to make them profitable. And there were conveyances that were invented without even a vague sense of the way society would construct their reality, automobiles and airplanes being prime examples.[2]

In the latter part of the 20th century, when California Governor Ronald Reagan called the supersonic transport (SST) "essential to America's continued leadership," or when New York Senator Daniel Patrick Moynihan spoke in similar terms about magnetic levitation (maglev), or when municipal officials insisted that high-tech light-rail vehicles (LRVs) were an essential badge of modernity—insisted on this even while neglecting prosaic modes of transit indispensable to the needs of the urban poor—they perpetuated a tradition.[3] Americans have often sought to link glamorous technologies to local pride or national honor or cosmic destiny, and to divorce them from constraints like market forces or, more importantly, from the obligation to allocate economic resources fairly. At the beginning of the 21st century, we hear demands that passenger trains be abandoned if they cannot hold their own ground financially, even though this is impossible elsewhere, even though national leaders seem to think nothing of throwing money at mismanaged airlines, and even though airliners use an infrastructure whose capital costs are distributed among the taxpaying public at large (as do barges and trucks). Could it have anything to do "with more politicians flying home from Washington than taking the train"?[4]

Rather than being isolated strands in the fabric of U.S. history, topics addressed here have been central to that history. Between 1815 and 1921, more than 30 million immigrants arrived, nearly every one of them aboard a ship. Improved modes of travel and transport were integral to the growth of market economies, industrialization, urbanization; today, a ubiquitous conveyance, the automobile, lies at the heart of our consumerist obsessions. The cultural significance of "the machine in the garden" likewise lies at the heart of historiographical debates that have not begun to lose their vitality. Newer but no less vital debates concern the role of consumers as agents of technological change, the nature of a place that Ruth Schwartz Cowan calls "the consumption junction." Even more important are questions about the extent to which "the relation between women and technology has diverged from that between men and technology." Consider that a ship is "she," consider that quite a few boys regard automobiles and airplanes not only as conveyances but also as toys.[5]

To return to evidence for the centrality of transportation to U.S. history, statewide canal systems in the 19th century and the interstate highway system in the 20th were public-works projects of unprecedented magnitude. Eastern trunk-line railways were the first billion-dollar corporations, and, as such, are assumed to have engendered complex new administrative hierarchies. To encourage construction in other parts of the country, Congress turned 131 million acres of public land (7 percent of the total land area of the contiguous states) over to private citizens, a fair number of whom were scoundrels. As early as 1860 a monopolistic New Jersey transportation company, the Camden & Amboy, was accused of "a studious concealment of its affairs" and at the turn of the 20th century the "Robber Barons" roiled populist resentment much like the masters of Enron and WorldCom at the turn of the 21st century.[6]

The development of technology was once portrayed as an unalloyed human benefit, though nobody can believe that any more. One may or may not agree with Lynn White Jr. that efforts to "control nature" by means of technological devices are a mark of Christian arrogance. But Americans were involved in devising "tools of empire," among them transportation technologies, to insure the global hegemony of the West, if only in the short term. After September 11, we know that complex technical systems can be redefined as "weapons of mass destruction." One may note that halting the SST in mid-air (so to speak) helped impart momentum to the American environmental movement in the 1970s. One must also note that Americans, who constitute 5 percent of the world's population, now drive nearly 30 percent of its motor vehicles and consume 65 percent of its energy, a shocking indulgence that imparts momentum to the "clash of civilizations" that has superseded the cold war as the dominant theme in world relations.[7]

And all this has been in consequence of laws that legislators have passed, or not passed; of steps taken by regulators, or not taken; of entrepreneurial urges held in check by conscience, or corrupted by power; of statesmanship or its failure. Steven Usselman writes that "many of the most tumultuous events in the nation's history have, at their core, involved disputes over the appropriateness and desirability of particular technologies," and what he has foremost in mind are transportation technologies. Responsibility for "internal improvements" was the most divisive political issue in the new nation, eventually exceeded only by the question of slavery. And, while the expression now sounds quaint, the issue still resonates as we ponder, for instance, the billions spent in completing a waterway in the 1980s that an Atlanta newspaper called "a 234-mile broken promise to the rural poor" and usually handles more bass boats than commercial barges.[8]

This booklet is an abbreviated address to an expansive subject. What I have sought to do is select stories pertinent to themes noted above, and to "make use of particularity to understand general problems and past experience." The history of travel and transport is blessed with contributions from giants of our literary heritage, including Richard Henry Dana, Francis Parkman, and Mark Twain. (Twain's very nom-de-plume was borrowed from riverboat lingo—when the leadsman at the bow called out "mark twain," the pilot knew he was in two fathoms [twelve feet] of water.) Accounts of the role of transportation in settlement and economic development such as George Rogers Taylor's *The Transportation Revolution* and William Cronon's *Nature's Metropolis* stand as landmark American narratives. Louis Hunter's *Steamboats on the Western Rivers* and John B. Rae's *The American Automobile* stand as classic analyses of the mutual interactions of technology, society, and culture. Recently, books like Carol Sheriff's *The Artificial River*, David Cecelsik's *The Waterman's Song*, and Kirkpatrick Sale's *The Fire of Genius* have begun to scuff the triumphalist veneer that suffuses much of the scholarship, as well as the popular literature. Both are vast, however, and if this booklet neglects a matter of special interest, please consult the endnotes and suggestions for further reading and reap the bounty provided by Clio, the muse of history.[9]

1

THE WESTERN OCEAN

Transportation technology has two components: there are conveyances, whether designed for water, dry land, steel rails, or blue sky, and there is infrastructure. For conveyances such as railway trains, canal boats, and highway vehicles, infrastructure of some sort must exist wherever they go. For those that ply the open sea, infrastructure is ordinarily confined to ports (or airports in the case of airways). The ocean itself is the highway. But to landlocked peoples an ocean looks like a moat. Such was the case historically with the Bering Sea, which blocked the advance of nomads who in the course of millenia had moved across northern Asia into Siberia. Beginning about 40,000 years ago, however—and for perhaps as long as 30,000 years—such peoples were able to cross over to North America on foot because much of the ocean had frozen into sheets of ice and the water level had receded to reveal a land bridge. By the time explorers arrived from across the Atlantic in the 15th century, there were perhaps 20 million residents of the Americas, scattered from Point Barrow to Cape Horn. South of the Rio Grande there were empires. To the north, many peoples were still nomadic, though not those encountered by the first Europeans to set foot on their lands. These natives knew that they occupied the edge of a vast continent. They may also have understood that the earth is spherical.[1]

Some Europeans understood. The cartographer Claudius Ptolemaeus, a Greek who lived in Alexandria during the second century C.E., had depicted the earth's configuration accurately, although he underestimated its circumference by 25 percent and overestimated the eastward extent of Asia. Such misapprehensions still held sway in the late 15th century; the prevailing European view of the earth was that it consisted of a tripartite land mass, with Europe and Africa separated from the coast of Asia by one great ocean, the Western Ocean. Norse navigators, notably Leif Erikson, had explored what they thought to be this coast from the 10th century to the 12th (actually, Erikson reached Newfoundland) but the open sea was a perilous place—especially for mariners without compasses—and the Norsemen tried to stay close to land. Not for another three centuries would mariners be enabled to set out directly westward from Portugal and Castile.[2]

POLITICS, TECHNOLOGY, AND EXPLORATION

The political impetus was provided by monarchs who put a premium on geographical discovery and the wealth and opportunities this might afford—first the Portuguese and then Spain under Ferdinand and Isabella, who united the kingdoms of Aragon and Castile and in 1492 destroyed the last Moslem stronghold on the Iberian Peninsula. Then there was the technological impetus, and this is a good place to take note of a scholarly debate. Because so much depends on decisions in political realms, there are those who deny any meaningful distinction between what is technological and what is political. For example, sociologist John Law lumps both together under the rubric "heterogeneous engineering," and this is a valuable counter to the notion that technology is capable of going places on its own. Yet, there is still that door: open or not? In the case of European exploration, the door had been opened by a set of devices, methods, and techniques.[3]

One of the devices was the compass, a Chinese invention that had diffused to Europe by the 13th century. There was also a method of steering vessels by means of a rudder hinged to the sternpost, an invention first evident on the North Sea at about the same time. Somewhat older was a technique of constructing vessels around a timber skeleton. Somewhat newer was an arrangement of sails in which some were laced to spars perpendicular to the masts (square rigged) and others were more or less parallel to the direction of travel (fore-and-aft rigged). A square rig was most effective with the wind astern, a fore-and-aft rig most effective for tacking. As Law puts it, a "mixed" rig—such as was common by the 15th century—could "convert winds from many directions into forward motion." With this capability, plus timbering strong enough to risk venturing into open seas, plus rudders and compasses, a door was open; absent these technologies, the New World would not have been explored as it was by seafaring peoples.[4]

Some tools were still missing or imperfect. Celestial navigation, a new science at the turn of the 16th century, made it possible to determine latitude (position on a north-south axis) from the altitude of the sun, corrected for declination on the basis of printed tables. But the available observational instruments, the quadrant and the astrolabe, were often useless aboard a ship tossing on stormy seas. Hence navigators had to be expert at dead reckoning, plotting a course and position from direction, time, and speed. Direction was determined with a compass and time measured by a half-hour glass, but speed could only be estimated. And not until the middle of the 18th century, with the development of the marine chronometer by John Harrison, was it possible to make accurate measurements of longitude at sea—that is, to determine how far one was from land to the east or west. Not until then was even the most skilled mariner not in danger of getting "lost at sea."[5]

Long before that, however, ships were being designed in essentially the same way they would be designed for hundreds of years. Though steering wheels would not replace tillers until the late 17th century, there were two or more decks and masts with a medley of sails. Such were the kind of ships available to Christopher Columbus, a Genoese who sailed due west from Palos in August 1492, having secured the patronage of Ferdinand and Isabella, and eventually reached a Caribbean island he named San Salvador. Columbus would make three more voyages across the Western Ocean, exploring ever more widely, yet he always misapprehended the earth's geography. He died in 1506, still under the assumption that somewhere beyond the lands he had visited there must be a strait, and then the Malay Peninsula, just as it appeared on maps derived from Ptolemy.[6]

But a Florentine adventurer, Amerigo Vespucci, had already referred to a *Mundus Novus* after claiming to have sailed to extreme southerly latitudes in 1501–02 and concluding that what he saw bore no resemblance to anything described in extant chronicles of Asia. A few years later a German geographer, Martin Waldseemüller, proposed that this New World be named for Amerigo, and in 1541 Gerardus Mercator prepared a map of the New World that had the letters **AME** across the northern part and **RICA** across the southern.[7] In the course of subsequent voyages of exploration during the 16th century, by mariners in the service of French, English, and Dutch regimes as well as Spanish, it became clear that this was another continent, or rather two of them connected by an isthmus, stretching close to both poles and separated from Asia by an ocean even more expansive than the Atlantic.

EUROPEAN SETTLEMENT

The Spanish were the first Europeans to plant a permanent settlement in North America, at St. Augustine in 1565. After the failure of several English ventures, the Virginia Company of London established a colony on the James River in 1607, and then in 1620 a 90-foot vessel called the *Mayflower* carried a group of religious dissenters who called themselves Puritans to the shores of Massachusetts Bay (Figure 1). One of them was William Bradford, who, in his *History of Plimmoth Plantation*, had this to say:

> Being thus arrived in a good harbor and brought safe to land, they fell upon their knees and blessed ye God of heaven, who had brought them over ye vast & furious ocean, and delivered them from all ye perils and miseries thereof, again to set feet on ye firm and stable earth, their proper element.[8]

Yorkshireman that he was, Bradford had a natural reverence for the "firm and stable earth," and the English Crown later granted the new settlers a considerable portion of it, a continental strip of land between 40 and 48 degrees north latitude—that is, from the head of Chesapeake Bay to the Gulf of St. Lawrence, nearly 1,000 miles. But except for passage afforded by rivers, this land would have been all but impenetrable because the entire eastern third of the continent was blanketed with dense forest. Native American trails could accommodate riders on horseback and pack animals, but not wheeled vehicles. For more than 200 years after the first European settlers braved the Western Ocean, America's rivers, along with coastal waterways such as Long Island Sound and Albemarle Sound, remained the only broad avenues of domestic trade.

Men of commercial mind, ever on the lookout for places where land and water met advantageously, found these prevalent along the coast of New England. In 1614, Captain John Smith wrote of sounding "about five and twentie excellent good Harbours, in many whereof there is anchorage for five hundred saile of ships of any burden; in some of them for one thousand."[9] Although Capt. Smith did not exaggerate, except for Boston and Salem most of the major Atlantic ports developed in more southerly latitudes, at New York, Philadelphia, Baltimore, Norfolk, Charleston, and Savannah.

In 1733 a London cartographer published a "Map of the British Empire in America with the French and Spanish Settlements adjacent thereto." The French controlled the continental heartland along with the most important rivers, the St. Lawrence and the Mississippi, until 1759 and 1804 respectively. As for the Pacific Coast, Spanish and British lands, while a Manila Galleon had occasionally called and there had been visits by English mariners, notably Francis Drake, cartographic data remained sketchy until the latter 18th century. Though good harbors were rare, one was magnificent. In 1776, Fr. Pedro Font, who accompanied the expedition charged with founding an Alta California mission honoring St. Francis, wrote that "the port of San Francisco is a marvel of nature, and might well be called the harbor of harbors." Before long, many Americans would be boasting about a "manifest destiny to overspread the continent" and in 1848 the harbor of harbors would be taken for the United States, along with 1.2 million square-miles of territory having only rudimentary avenues of transport and travel. San Francisco was already a busy destination for interoceanic commerce and a portal to "the Orient." Now, it also looked like the destination for a railway that would "redeem a wilderness from savages."[10]

Figure 1. This Smithsonian Institution model of the *Mayflower* is conjectural in a few details but accurate in its essentials. Square-rigged on two masts with a fore-and-aft sail on a third—the nautical term is bark—the *Mayflower* was typical of small English merchant vessels of the early 17th century.

2

A NEW NATION

The British Crown sought to dominate the economic interests of its North American colonies by enforcing a doctrine known as mercantilism. In theory, colonists were constrained to ship raw materials to England, and only to England, in exchange for processed and manufactured goods, necessities and luxuries. In practice, colonists often managed to evade strictures on "enumerated" goods, and the Crown tolerated and eventually even encouraged certain industries. Foremost was shipbuilding. The colonies were rich in "naval stores"—rosin from pine gum, turpentine from pitch, hardwoods for masts, spars, timbers, and planking. The white pine of New England grew to magnificent dimensions, trees 200 feet tall with 9-foot girths, and the availability of such timber for masts facilitated the construction of ships larger than ever before. By the time of the Revolution, tonnage built in North America accounted for a third of the British fleet.[1]

Besides naval stores, commodities actively traded included foodstuffs from the middle colonies; cotton, indigo, and tobacco from the south; and sugar from the Caribbean, the latter usually converted into something more transportable, rum. In the "triangular trade," sugar from the Caribbean was distilled in New England and then shipped across the Atlantic to Africa and exchanged for slaves. The colonists also traded with the Native Americans for furs. As tribes were decimated by warfare and disease, however, and European settlers took control of land further away from the coastline, river transport assumed increasing import. When the colonies declared their independence, small fore-and-aft-rigged single-masters called sloops, running more or less on schedule, linked every American port from Maine to Georgia and sailed the lower reaches of the major rivers. There was also a road system, of sorts.

ROADWAYS

At the end of the 18th century, the population of the United States was 3.9 million, 700,000 of whom were held in bondage. It was almost evenly divided between north and south, with a few more than 200,000 beyond the Alleghenies and 200,000 in urban areas. Urbanization was a slow process at first, and the decade between 1810 and 1820 saw a negligible shift from country to city. Among the seaports only Baltimore exhibited anything like rapid growth. But the inauguration of steamboat traffic on the Ohio and Mississippi in 1812—the same vessel that took goods downstream could now return against the current in a reasonable period of time—had an immediate impact on Cincinnati and New Orleans. By 1820, 2.4 million people were living in the Mississippi Valley, many of them increasingly engaged in a market economy as a result of declining transportation costs.[2]

Even before the advent of steamboats, the Ohio and Mississippi Rivers had afforded swift passage in one direction, southerly (just as the Atlantic's prevailing winds and currents made for a great "downhill" run eastbound), but overland travel was entirely different. While there were post roads, of course (Boston and New York were connected by 1673), pack animals required only a trail, and trails had been well established by Native Americans. Four-wheeled vehicles needed more than a trail. Like sailing ships, such wheeled vehicles incorporated a combination of technological devices, the fundamental one naturally being the wheel itself, the spoked variety dating from 2000 B.C.E. Iron tires and the use of tallow lubricants were ancient technologies as well. The horse harness was more recent, as were nailed iron horseshoes and the whippletree, which equalized the pull when a conveyance made a turn. With a well-crafted vehicle, one horse could pull 10 times more weight than it could carry on its back. Conestoga wagons, sometimes drawn by teams of six horses, could be used to supply an army.[3]

But infrastructure was weak. This was particularly so in the vicinity of coastal wetlands and on the rising terrain beyond the Piedmont. During the warfare that led to cession of French territory east of the Mississippi in 1763, military roads had been cut through the forests of the Alleghenies and on to the "forks of the Ohio" (Pittsburgh). After the Revolution, however, road-building was ordinarily left up to privately chartered companies, the first of note being the 66-mile Philadelphia and Lancaster Turnpike, completed in 1794. Here, teamsters operated heavy wagons, though most over-the-road transport involved vehicles of considerable "elasticity"—not the last time, as Carroll Pursell points out, "that American technology was to be characterized as 'light and flexible.'"[4] This elasticity was essential wherever infrastructure remained rudimentary, which was actually almost everywhere. For "improved" roads through difficult terrain—roads that were graded for drainage and surfaced (and thus passable in all seasons)—only the national government seemed to have the wherewithal. But it was with plans for developing such roads that politics intruded upon the history of American transportation in an entirely new way. A door was open, yes, but not everybody thought it prudent to step through.

The founders of the Republic were well aware that a territory the size of the United States had not been unified since the time of the Roman Empire. They also knew that the Romans had put a premium on efficient transport, and in 1807 Treasury Secretary Albert Gallatin unveiled a proposal for binding together all the states and territories. It was, as Ruth Schwartz Cowan put it, a "transportation master plan for the new country."[5] In the mid-20th century, such a proposal could seem unexceptionable—by the time President Eisenhower signed the Federal-Aid Highway Act in 1956 (the initial appropriation was $25 billion) dissent had all but vanished—but in an earlier era this became the most contentious of political issues. At the heart of the issue was the Federal Constitution, and disagreement about the extent to which the phrase "necessary and proper" could be interpreted as authorizing internal improvements. The private sector lacked financial power, as did most states. When urban elites had money to invest, they typically elected to back mercantile ventures. That left the central government. "Loose constructionists" under the leadership of Alexander Hamilton and the Federalists believed that the Constitution authorized internal improvements; "strict constructionists" under the leadership of James Madison did not.

Neither did Thomas Jefferson, not initially. But after his election to the presidency in 1800 Jefferson took several steps not explicitly sanctioned in the Constitution, notably the purchase of a vast parcel of land that almost doubled the size of the nation and the financing of an improved road from Cumberland, Maryland (from a connection with a road from Baltimore), over the mountains to the head of navigation on the Ohio River, and beyond. This was the Cumberland Road, later called the National Road (Figure 2). Begun in 1811, the initial segment was 30 feet wide, with ditches alongside and large foundation stones underlaying gravel paving, a technique borrowed from the Romans. Although the National Road was eventually completed to Vandalia, Illinois, the national government's responsibility for such projects was never accepted by strict constructionists, who represented a political stance that remains compelling to this day—aversion to public debt and suspicion of centralized authority.[6]

Figure 2. This 1820 scene on the National Road, with ox-drawn Conestoga wagons, suggests that lofty standards of construction had slipped by the time the road reached Indiana. Conestoga wagons were first made by settlers of German origin along the Conestoga River in southeastern Pennsylvania in the mid-17th century. The bowed configuration later gave rise to the nickname "Prairie Schooner." Courtesy Indiana Historical Society Library.

During the latter 19th century, a deficient infrastructure would give rise to a crusade for "good roads," and eventually a consensus about these being a national imperative with a strong element of federal responsibility. Earlier that consensus was elusive, however. Even though Gallatin's plan was eventually realized in large measure, it took the form of partnerships involving the national government with state governments and with private entrepreneurs who were granted special privileges such as loan guarantees and land grants. The nature of these partnerships was not always explicit. For the purpose of funding turnpikes, as private toll roads were called, many states passed legislation authorizing joint stock companies (the Philadelphia and Lancaster was the first), thereby altering the American financial landscape. Often this legislation permitted state governments to use tax monies to buy stock, in effect subsidizing private enterprise.

Costs were always an issue, and most turnpikes were laid out with a minimum of grading, a constraint often specified in their charters. Surveyors or "mearsmen" worked with a compass or a circumferenter, an instrument for measuring horizontal angles. All labor was hand labor, the only assist coming from explosives. Lacking dynamite and time fuses, workers used ordinary black powder for blasting. The need for vast numbers of picks and shovels provided a major stimulus to local iron industries: for example, the Ames Shovel Works in Easton, Massachusetts, got its start in 1804 when Oliver Ames developed techniques for making a dozen shovels at a time, sort of a proto-assembly line. Turnpikes were sometimes surfaced with wood, especially in marshy areas. The most common technique consisted of laying small, closely spaced tree trunks crossways to the direction of travel. These were (naturally enough) called corduroy or washboard roads, and one such road linked Newark and Jersey City for 85 years. Roads were also surfaced with planks spiked to stringers, a technique originating in Russia as early as the 10th century; or using another Russian technique, hexagonal woodblocks set close together on end; or even with charcoal, wood pulp, and sawdust.[7]

Wood deteriorated and wore out quickly and no wooden road was ever very satisfactory, but certainly it was preferable to no surfacing at all. The best surface was called Macadam, named after a Scotsman who came up with the idea of covering a graded substrate with broken stone, ideally granite, and compacting it with heavy wheeled vehicles to give the cross-section the contour of an arch and facilitate drainage. Hardly less than wooden roads, the success of Macadam depended on the skills and honesty of contractors and an attention to maintenance that proved difficult to sustain. This was doubly true with bridges, but here even fundamental knowledge was lacking at first.[8]

Novelty

In colonial America, bridges amounting to more than a row of wooden stringers were rare. Streams usually had to be forded, and entrepreneurs established ferries at major crossings. Yet there were stone-arch bridges in Europe dating to the Roman Empire. In 1779, the Shropshire ironmaster Abraham Darby built a 196-foot cast-iron bridge across the Severn near Shrewsbury, the first in the West, though anticipated in China. While Americans knew about the iron bridge over the Severn, of the requisite technical skills for such an undertaking there were none on their side of the Atlantic for many years. "Public spirit is alone wanting to make us the greatest nation on earth," wrote one enthusiast in 1811—who continued that nothing was "more essential to the establishment of that greatness than the building of Bridges"—but in truth a lot more was needed than public spirit. Not until the late 1830s would there be an iron bridge erected in the United States and massive stone arches did not begin to appear until even later than that.[9]

Wood was another matter. Even by the time of the Revolution, there were long wooden bridges that rested on wood pilings, a rarity in Europe because of deforestation. A bridge spanning the Charles River between Boston and Charlestown was completed in 1787, another across the Schuykill at the eastern end of the Lancaster Turnpike a few years later. Then, in the early years of the 19th century a number of wooden *truss* bridges were erected in New England, one of the first signs of a technology that seemed peculiarly American; indeed, bridges covered with planks to protect the structure from the elements became an American icon. Truss bridges neither support the roadway from below, as in an arch or trestle, nor from above, as in a suspension bridge. Rather, they are designed to stiffen the roadway itself by means of an interconnected arrangement of triangular members, some in tension, some in compression. A triangle is geometrically rigid. Systematic knowledge of trusses had represented an open door since the Renaissance, and in Europe trusses often supported the roof and ceiling of churches. But truss *bridges*, as Barrie Trinder puts it, were "particularly suited to North American conditions."[10]

Wood was plentiful, as were woodworkers with skills learned in shipbuilding, as were mental images of trusses configured in various sorts of ways. Reflecting a special American enthusiasm for patents of invention, designs became proprietary. In 1820, Ithiel Town, a New Haven architect, patented a rather simple truss and then sold rights to bridge builders at the rate of a dollar a foot. With no basis in mathematical calculations, the durability of Town bridges was assured by the expedient of overbuilding—using such a dense latticework that failure seemed unlikely. Steven H. Long, a man of many parts (he also designed railway locomotives in Philadelphia and took steamboats up the Missouri River as far as Fort Benton, Montana), was the first American to devise a truss on the basis of abstract calculations of stress. Long's design was later refined by William Howe, who patented a truss having wooden diagonals and vertical wrought-iron tension rods with turnbuckles to adjust for sag. Thomas and Caleb Pratt's similar but more refined design was utilized with many early railway bridges.[11] (See Figure 3.)

For railways with the heaviest traffic, trusses combining wood and cast-iron members soon gave way to cast-iron exclusively. But these were subject to catastrophic failure and eventually gave way to wrought-iron and finally steel; the first steel bridge spanned the Mississippi at St. Louis in 1874 and was named for its engineer, James B. Eads. Stone arches became common in railway construction, and wooden trestles were ubiquitous, often consuming dazzling quantities of lumber (Figure 4). Lucin Cutoff, a 1903 realignment of the first transcontinental railway, stretched 32 miles across Great Salt Lake, 11 miles of which rested on thousands of wooden piles driven deep into the lake bed.[12]

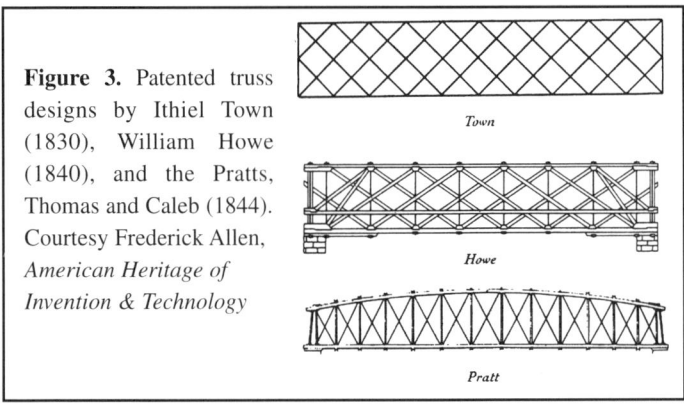

Figure 3. Patented truss designs by Ithiel Town (1830), William Howe (1840), and the Pratts, Thomas and Caleb (1844). Courtesy Frederick Allen, *American Heritage of Invention & Technology*

EXPANSIONISM

"Bridges define our landscape," writes David McCullough. They represent "beauty, power, and rational accomplishment," adds William L. Richter. In the United States during the early 19th century, however, roadways often represented a lot of controversy. Gallatin theorized that federal initiatives would be constitutional if funded by the sale of public lands, and in the 1820s Henry Clay gave that vision a boost with his "American System" of planned development. But Andrew Jackson knew as well as anybody that no technology is ever promoted without some people losing as others gain, and in 1830 he vetoed legislation authorizing improvements to a road connecting Lexington, Kentucky, with Maysville, on the Ohio River. Historians often denote this veto as a turning point in federal support of internal improvements, especially since Jackson's successor, New Yorker Martin Van Buren, was hostile to any project that might threaten the hegemony of the Erie Canal, financed with state money. Actually, the national government, "Washington," continued to finance military roads in the territories—along the 94th meridian connecting forts of the "Permanent Indian Frontier" and elsewhere—and after the Mexican War the Pacific Wagon Roads Office built and

Figure 4. Trestlework often preceded an earthen fill, which served as fire protection (trestles were notoriously combustible). Here, filling has begun even while a trestle is still under construction. The site is the Papillion Valley near Omaha, a 1906 realignment of the original line of the Union Pacific. J. E. Stimson photograph, courtesy Wyoming State Archives, Museums and Historical Department.

maintained roads as well. More important, the Army Corps of Engineers got annual funding to improve conditions for navigating rivers and harbors, and of course western railways were subsidized in lavish measure.[13]

The vast expansion of American territory during the Polk administration—the Oregon Treaty in 1845, the Mexican Cession three years later—left the nation in a disconnected state similar to Jefferson's time, when "the west" had meant west of the Alleghenies. After 1848, it meant something else: a semiarid expanse later dubbed the Great American Desert, then mountains far more precipitous than the Alleghenies, then more mountains, and finally another ocean 3,000 miles from the Atlantic. Until 1855, the Pacific Coast was reached most conveniently (if the word makes any sense in this context) by a voyage around Cape Horn, some 16,000 miles. After that, travelers had an option of making two separate voyages punctuated by a train trip across the Isthmus of Panama. Then, after 1869, San Francisco could be reached directly by railway.[14]

But that is getting ahead of the story. Road or rail? Rail or canal? Sometimes Americans confronted more than one open door and choices were driven by political considerations—if not necessarily rationality—with every contingency being "potentially malleable." Some scholars believe that "technical" factors are invariably susceptible to "sociological explanation," but that cannot be. In the case of European exploration during the 15th and 16th centuries, there was a set of technological contingencies that do not appear to have been malleable, and for oceanic transport and travel in the 19th century they still were not. A vessel had to be "seaworthy," capable of functioning in a dangerous, not to say hostile, environment—though when it came to actual design the imagination could soar, and it did, yielding novel conveyances of great dramatic power.[15]

3

SEAWAYS

After the Revolution, Americans struggled to establish the right to engage in international trade in a fiercely competitive world that scarcely believed the achievement of independence. Excluded from familiar commerce, merchants were impelled into a search for new markets. Most notable was China, a market formerly monopolized by the British East India Company. China's prime commodities were luxuries (silk and the porcelain tableware that became known as chinaware, or simply china), and a perishable staple, tea. Eventually, millions of pounds of tea were imported for domestic consumption or re-export; New York, Salem, Boston, and Providence tea fortunes provided capital for many of the nation's early ventures in transportation and industry. By 1807, the annual value of all imports and exports was well over $100 million each. By mid-century both imports and exports were ranging towards $300 million. This implied busy seaways.[1]

Prosperity in foreign trade was intimately linked to speed, and here the Americans had already hastened through an open door. In the years just after independence, speed was the first line of defense for a merchant marine that had no protective navy but was often getting tangled in foreign wars. And even before 1776, smugglers who defied the decrees of British mercantilism had required the most elusive of vessels. Most notable was a type of schooner (the term designates a vessel with two masts, fore-and-aft rigged) built in small yards tucked along the shores of Chesapeake Bay, the first distinctly American watercraft. Their heyday came during the Napoleonic Wars and the War of 1812, but many remained active in illegal trades through the 1830s. Slave ships represented a technological adaptation to a peculiar trade. They were heavily sparred—a large expanse of sail enabling speed even in light breezes—but lightly timbered. Their cargo, though valuable, was not heavy.[2]

At the same time, larger and more traditionally designed ships (in technical parlance, a ship is a square-rigger with three or more masts) were being launched in port cities north of Chesapeake Bay. Most important were the packets, so called because their cargo included packets of mail. After a stable peace was finally established in 1815, entrepreneurs used packet ships to launch a revolution in transatlantic commerce with the establishment of regular schedules. Always before, travelers waited for passage on vessels that would depart only when fully laden with cargo. Beginning in January 1818, however, ships operated by a firm called the Black Ball Line set sail from New York for Liverpool precisely as announced in advance —"on their Appointed Days, full or not full." This proved to be a convenience for which people would pay handsomely, and by the mid-1840s there were more than 50 trans-Atlantic packets sailing out of New York alone.[3]

POLITICS, TECHNOLOGY, AND GOVERNMENT PATRONAGE

But there was also serious competition of a different sort. Even though a packet line might guarantee a time of departure, the concept of a definite schedule did not apply to arrivals, especially on a westward voyage which typically involved adverse winds and constant tacking, and could make such a trip 500 miles longer than an eastbound passage. By the 1840s, British-flag steamships, especially those operated by Samuel Cunard, a Canadian, were undermining the preeminence of the sailing packets on trans-Atlantic routes. As designed by the Scottish engineer Robert Napier, they were powered by means of paddlewheels located amidships on either side.[4]

Even though Americans had experimented with auxiliary steam power as early as 1819 aboard a ship named the *Savannah* (Figure 5), this was only a foot in a doorway that was not yet open. Steam power was well established on American rivers, but firewood was still plentiful on nearby shores and bunkers could be refilled whenever necessary. Not so on the high seas. Not until nearly 20 years later was a ship enabled to cross the Atlantic entirely under steam, and it was British. More such "teakettles" followed, larger than the sailing packets and faster. These soon drew off first-class passengers and "fancy" manufactures, leaving the sailing packets only with immigrant traffic and bulky cargoes, principally cotton, which topped 50 percent of American exports by mid-century.[5]

Figure 5. The *Savannah* is depicted in this drawing published in the *Annual Report of National Museum for 1890*. She was intended for coastwise service, but a business slump induced the owners to try and sell her abroad. She thereby became the first ship equipped with an engine to cross the Atlantic, though steamed only for 90 hours during a 29-day voyage to Liverpool in spring 1819.

By then, Cunard's British and North American Royal Mail Steam Packet Company was scheduling 44 transatlantic roundtrips annually. The name pointed to Cunard's success at following the money: profitability was tied to an annual mail subsidy (60,000 pounds-sterling at first) from the British government. Naturally, Americans pressed for a similar privilege, a contract for carrying mail abroad at premium rates. The premier route was New York–Liverpool, and the prize eventually went to Edward K.

Figure 6. The *Baltic* had two engines rated at 1,100 horsepower each, which required huge bunkers and minimized cargo capacity. Even though propellers were in use on the Great Lakes by the 1840s, operators of oceanic steamships preferred paddlewheels until the 1850s. Lithograph by Nathaniel Currier, courtesy National Museum of American History.

Collins, president of one of the trans-Atlantic packet lines. In 1847, Congress awarded Collins's New York and Liverpool United States Mail Steamship Company a $385,000 subsidy to make 20 roundtrips a year with a fleet of five ships.[6]

Collins ordered four of them at once, the *Atlantic*, *Pacific*, *Arctic*, and *Baltic*, and these marked another notable innovation (Figure 6). They were sidewheelers, like Cunard's, but, with engines working at half-again as much pressure, they were capable of better than 13 knots. The *Baltic* made a crossing in 9 days, 17 hours, topping the best Cunard time by several hours. This may not have made any difference in rational terms, but—as software marketeers prove over and over in our own time—people are fascinated by speed. Americans, New Yorkers especially, treated the Collins-Cunard competition as a dramatic sporting rivalry.[7]

Though technologically bold, the Collins liners were at the mercy of political patronage. They could carry nearly twice as many passengers as Cunard liners, but they were far more costly to operate and actually had *less* cargo capacity: those huge engines consumed a lot of fuel. Collins might have considered economizing but instead he went in quest of a larger subsidy, appearing before Congress wrapped in the flag of national honor. A political uproar ensued, opponents contending that he had squandered his initial sum by chasing frivolous speed records. Collins did manage to get his subsidy more than doubled, to $858,000, but with a proviso that it could be terminated at any time.

Catastrophe loomed on the high seas. The *Arctic* sank after a collision in 1854 (among those who perished were Collins's own wife and daughter), and, in the enigmatic expression of mariners, the *Pacific* "went missing" two years later, probably having struck an iceberg, perhaps running too fast in dangerous waters. The combined death toll was more than 500. Rumors spread about the dangers inherent in subjecting wooden ships to the stress of throbbing engines. There were many close calls with icebergs. As a replacement for his lost ships, Collins bought the *Ericsson*—a curiosity when originally outfitted with hot air ("caloric") engines, though now converted to steam—but she was too slow to meet the terms of his contract.[8]

For Americans who pictured the North Atlantic as a "great colossal raceway," the status of the Collins Line plunged when Cunard's new *Persia*, made of iron, cut more than 16 hours off the record. Collins rushed completion of the *Adriatic* at a cost of more than a million dollars. She was huge, 344 feet long with 376 passenger spaces, and incredibly fast. With engines from New York's Novelty Iron Works and a hull designed by George Steers, who also designed the *America*—with which the United States had first established preeminence in international yacht racing—she could make 18 knots. But she was terribly costly to operate and this precluded any challenge for the trans-Atlantic speed record. Her maiden voyage in 1857 took place amidst a financial panic and a deepening sectional crisis. She was nearly empty. Congressional enthusiasm for a "transatlantic postage tax" had evaporated and political disfavor spelled doom. With the company $750,000 in debt, its three ships brought a total of $50,000 at auction.[9]

CLIPPER SHIPS

Another story played out on the Pacific, equally dramatic. Even as steamships were clearly the wave of the future on the Atlantic by the 1840s, on interoceanic trade routes around Cape Horn steamers were unable to carry sufficient fuel. But experience with swift privateers had shown another door to be open, sailing ships designed for ultimate speed. The word *clipper* initially denoted a fast racehorse, but it ultimately belonged to square-riggers built in the 1840s and 1850s for passage to China, California, Australia—wherever distances were long and quick passage commanded premium rates. This epoch lasted as long as gold-seekers rushed to San Francisco and Sydney and as China traders pounced on a newly opened opportunity to ship tea directly into English ports, about a decade.

Clipper ships were two and three times longer than had been prevalent with sail, and they were sparred for vast spreads of canvas such as had never been seen before. As with so many conveyances, there was no one "inventor," though a discrete epoch began in 1845 with the completion at East Boston of the *Rainbow*. She was the design of a young naval architect named John Willis Griffiths, who had apprenticed with William H. Webb, the preeminent New York designer whose 150 ships included 15 packets for the Black Ball Line. Griffiths's next creation, called the *Sea Witch*, carried some of *Rainbow*'s characteristics to extreme—concave bow-lines and billows of canvas including "skysails" atop the usual rig and "studdingsails" set by extending the yards with booms—and she set one record after another. Then, there was Donald McKay, dubbed the "Michelangelo of shipbuilders." Ordinary merchant ships usually required 20 or 30 weeks to make the 16,000-mile voyage from New York to San Francisco; McKay's *Flying Cloud* made it in well under 13 weeks and his brother Lauchlin once drove another clipper on a phenomenal South Pacific reach of 1,478 miles in four days.[10]

Never before had any kind of human conveyance been capable of going so far so fast. Yet, in their own way clipper ships were as impractical as the Collins sidewheelers. Because of their lean configuration, clippers could not carry much cargo. Driving them to record-setting passages required more than a hundred crewmen and working aloft was incredibly dangerous (Figure 7). In a different way, so was investment in clipper ships. By 1855 there were too many ships and too little business, and the depression that began in 1857 amounted to a death-knell for "greyhounds" on long around-the-Horn routes.[11]

After the Civil War, American shipyards continued turning out full-rigged wooden ships that were not so lean and lofty but far more economical. Many came from Maine, hence the generic Down Easter. While these rarely if ever covered their initial cost in one voyage (as a few California and China clippers had done at mid-century) they managed to turn a profit on interoceanic routes for many years. Cargoes were nothing if not prosaic: coal, lumber, sand, bricks, stone, salt, and ice, grain, salt fish, and livestock. Peruvian guano became especially important. By the time the last wooden square-rigger was launched in 1893, it was clear that the United States had slipped from the oceanic vanguard. Compound engines (with one cylinder exhausting into another), steel hulls, and screw propellers had become the state of the art, and such ships

Figure 7. The *Great Republic*, designed by Donald McKay and launched in 1853, was the last of McKay's several merchant vessels to be designated the largest afloat. Among her 120 crewmen, some had the daunting task of climbing to the top of the mainmasts and tending to the double topsails. From a French lithograph, courtesy National Museum of American History.

Figure 8. These seven men, posing in 1917 with their mascot, represented an average-sized crew for a coastal schooner in the last days of such vessels. Courtesy The Mariners' Museum, Newport News, Virginia.

were rarely built for the U.S. merchant marine. What did get built were wooden schooners for coastwise traffic, particularly coal from the Virginia tidewater and lumber on the Pacific. A fore-and-aft rig could be worked upwind more easily than a square rig, sail being handled by just a few men, from the deck (Figure 8). Yet, even here designers pushed the limits in order to minimize labor costs. In a saga that also played out with railway locomotives, schooners reached extravagant dimensions—several with six masts and one with seven, the *Thomas W. Lawson*—before commercial sailing ships disappeared from the American transportation panorama once and for all.[12] (See Figure 9.)

Figure 9. In their quest "to move the most cargo with the smallest payroll," coast-wise shippers began ordering four-masted schooners in 1880, then five-, six-, and finally in 1902 a colossal seven-masted schooner 395 feet long, the *Thomas W. Lawson*. Fully laden, she could not get in and out of East Coast harbors and she spent most of her days in tow between the Gulf and Delaware River refineries. Finally, in late 1907 she was chartered to carry 6,000 tons of crude oil across the Atlantic. On Friday, December 13, she went aground and broke up on one of the Scilly Islands, 25 miles from Land's End, Cornwall. The sea claimed all but two of her men as well as her entire cargo, the first major marine oil spill, and at a spot not far from one of the most notorious, the *Torrey Canyon* disaster 60 years later. Courtesy National Museum of American History.

4

CANALS

As with roads, the technology of canal building was ancient: "as old as Ramses of Egypt and Xerxes of Persia," writes Dirk Struik. A waterway connected the Mediterranean and the Red Sea thousands of years before the Suez Canal, and archaeologists have discovered remains of canals between Lake Okeechobee and the Gulf of Mexico that date from the third century C.E. As a technique for raising and lowering vessels to different elevations, the gated lock chamber dates from the 13th century in Ming China, on the Grand Canal linking Beijing and Hangzhow. In the West, the Canal du Midi in France was thriving before 1700, the Bridgewater Canal in England by the 1760s.[1]

In North America, there were promising sites for canals at points along the coast where a "short cut" beckoned—across the shoulder of Cape Cod, across New Jersey from the Hudson River to the Delaware, across the Delmarva Peninsula separating Chesapeake Bay (and Baltimore) from the Delaware River (and Philadelphia), across the neck of swampland separating Norfolk from Albemarle Sound. Beginning in the 1780s, several diversions were completed that skirted rapids in rivers, the first providing navigation around the falls of the James River at Richmond.[2]

As with iron bridges, however, when it came to artificial watercourses that required locks because of changing elevation, something essential was lacking—know-how. Beyond the challenges of designing a dam that would enable filling a canal from a nearby lake or river, or a retaining wall that would stand firm, there were others, much more difficult if locks were needed. How to bond bricks or stones lining a lock chamber and prevent "weeping"? How to design movable gates to withstand tons of water pressure? And how to manage a large workforce among whom unrest was endemic? David Cecelsik describes canal building as "the cruelest, most dangerous, unhealthy, and exhausting labor in the American South," and it would have not been much different in New England.[3]

The first notable American canal was the Middlesex, 27 miles, linking the Charles River at Boston and the Merrimack at Chelmsford. The superintendent of construction, Loammi Baldwin, had experience as a surveyor but had never seen a canal lock. As costs escalated dangerously, he had to seek advice from an Englishman, William Weston, who immediately discovered that the original surveys for ascent and descent were awry. The first lock "broak and failed" as soon as it was filled with water. After nearly a decade, the canal finally opened in 1803 at a cost of more than a half-million dollars, a huge sum for the day. There were 8 aqueducts, 20 locks, and 48 bridges, all in all the most elaborate construction project yet completed in the new nation. Disappointing returns discouraged other projects of the same sort in New England for two decades. Enthusiasm would revive, but only briefly, for railways provided devastating competition. The Middlesex Canal was out of service by 1852, crushed by the Boston & Lowell Railroad.[4]

THE BIG DITCH

After the National Road proved to be so contested, it seemed easier to marshal support for internal improvements at the state level. Pennsylvania and Maryland assayed ambitious canal projects, but much of the terrain there was difficult. On the northern flank of the Alleghenies in New York state, however, there was a low-elevation route between the Hudson River and Lake Erie via the Mohawk Valley. Boosters argued that a canal would enrich the entire state, hence New York's taxpayers should assume the bonded debt.[5]

Chief among these boosters was De Witt Clinton, who had served in the Senate and as mayor of New York, and even run for president before his election as governor in 1817. Clinton immediately launched a campaign to attract financing, and ground was broken for the first segment, the easy 75 miles between Utica and the Seneca River, on July 4, 1817. Its quick completion yielded funding for the more difficult stretches, which involved 18 aqueducts over streams and 84 locks to lift traffic 630 feet above the level of the Hudson and then, at Lockport, 62 feet down from the Niagara Escarpment to the level of Lake Erie. Hardly less so than Loammi Baldwin, the men in charge of construction, Benjamin Wright, Canvass White, and James Geddes, entered the fray as amateurs. But they knew where to go for expert help. They also benefitted from the development of two fearsome inventions, one that snapped large trees off rather than cutting them down, another that could pluck whatever remained from the ground like turnips. Luck played a role, too, with the discovery in Herkimer County of rock deposits called trass, ideal for making the "hydraulic" cement essential to the durability of stone locks that would be submerged.[6]

Clinton was absent from the statehouse for three years after 1822, but when he returned it was as a hero who had surmounted one political obstacle after another to complete the greatest canal in the world, a 383-mile link between the Great Lakes and the Atlantic. And that was taking no credit from Wright, White, and Geddes, who had earned the admiration of European engineers thoroughly conversant with canal construction. In retrospect, the Erie came to be regarded as an American "school of general engineering" whose lessons were applied to many other projects scattered far and wide.[7]

While not the longest canal in the United States, in every way the Erie was outstanding; in the early 1840s, a European visitor wrote that the nation's burgeoning transportation system in its entirety was "based on the success of the Erie Canal." The longest American canal was the 450-mile $8.2-million Wabash and Erie, running from Evansville, Indiana, on the Kentucky River to Lake Erie at Toledo. On paper, this canal looked like a way to "get out" of Lake Erie, just as the Erie Canal was a way to "get in." But the traffic was just not there, and needed repairs went unfunded. On the other hand, the Erie paid off its $7.9-million cost from tolls after only 10 years and then was enlarged from its 4-foot depth and 40-foot width to 7 by 70. It was again enlarged in the 1880s, and portions of it survive even today as the New York State Barge Canal.[8]

Although this canal now carries mainly pleasure boats, not barges, the Erie remained commercially significant for 150 years. Its best year came in 1880 when it handled 4.6 million tons of freight. After that, the New York Central Railroad, with its "water-level route" between Albany and Buffalo, was a relentless rival. Yet canal tonnage topped four million twice in the 1930s and only occasionally before the 1970s slipped below a million. The Erie Canal insured that New York City would remain the nation's preeminent entrepôt, not Boston or Philadelphia or Baltimore. Before it opened, moving a ton of grain from Buffalo to New York cost $100; after 1825 the cost was $9.[9]

Mania

Nothing succeeds like success, and during the second quarter of the 19th century several major canals were completed in New England and the mid-Atlantic region, including the Chesapeake and Ohio (C&O), which linked Cumberland, Maryland, with Georgetown, and, for a time, with the Potomac River via an inclined plane. The C&O was an expensive undertaking, but not nearly as expensive as the Pennsylvania Main Line, connecting Philadelphia and Pittsburgh. The engineer in charge, Moncure Robinson, had to confront the crest of the Alleghenies, which he surmounted by means of a Portage Railroad between Hollidaysburg and Johnstown—a 37-mile series of inclined planes that carried loaded canal boats to 2,200 feet (Figure 12). This was more than three times the Erie's highest elevation, and the Main

Figure 12. On the Pennsylvania Portage Railroad, fully loaded canal boats were hauled over the crest of the Alleghenies by means of stationary steam engines. Courtesy National Archives.

Line's 174 locks were more than twice as many as on the Erie. Like so many other sagas in the history of American transportation, it was a triumph largely in a "technical" sense. The Main Line cost $10 million, and yet this was only a fraction of what Pennsylvania—quite unsuited to this mode of transport topographically—eventually spent on canals. Virtually all this money, reported by one historian to be in excess of $100 million, had to be written off with the advent of railroads.[10]

Between 1790 and 1850, 4,400 miles of artificial waterways were created in the United States (Figure 13). Not much survives. Some canals were poorly engineered, others had disadvantageous routes, many were mismanaged, nearly all suffered fatally from railway competition. After 1840, few of them were able to cover operating expenses. Pennsylvania was not the only state left in dire straits by canal mania. Sparked by dreams of tapping into eastern markets for farm produce, Ohio and Indiana borrowed lavishly to finance canals. With an annual budget of only $50,000, Indiana floated $10 million worth of bonds; Indiana was one of several states that later changed their constitutions to outlaw state investment in transportation ventures.[11]

The few canals that remained in private hands after the Civil War became government property by default. Among the hardiest of these was the C&O, a coal-hauler which lasted until 1924 despite getting battered by floods and despite the proximity of the Baltimore & Ohio Railroad, which actually did reach Ohio, unlike the canal, and by 1874 reached Chicago. A portion of the C&O near Washington, D.C., still serves as a tourist attraction, as is the case in New York, and also in eastern Pennsylvania, a portion of the once-extensive system of anthracite canals, the Morris Canal. Canals with continuing commercial import include the Chesapeake & Delaware, Albemarle & Chesapeake, and the Cape Cod, none having any locks and each affording passage to oceangoing ships. By and large, however, the infrastructure of the American canal system has now disappeared altogether or is discernable only to the keen-eyed industrial archaeologist.[12]

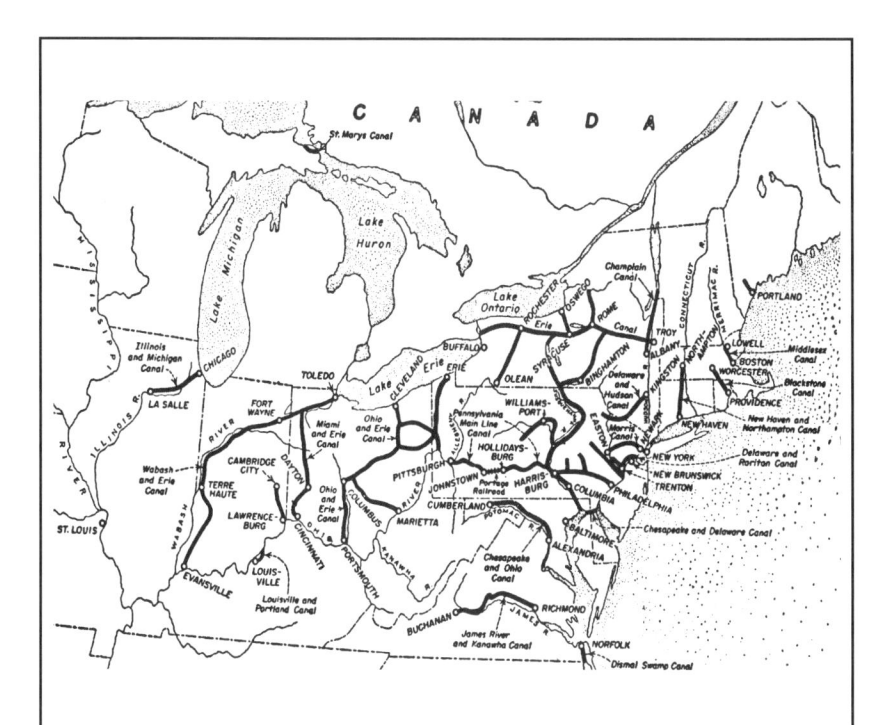

Figure 13. The principal U.S. canals in 1850, several of which would fail within a few years. From George Rogers Taylor, *The Transportation Revolution, 1815–1850* (New York: Harper and Row, 1951).

5

AN INLAND SEACOAST?

The Great Lakes constitute 95,000 square miles of waterway amidst what was long one of the world's great natural bonanzas. After completion of the Erie Canal, settlers poured into the region. The commercial mainstay was grain and Buffalo became the capital of the milling industry. Then, in 1855 traffic began to develop between ore fields on Lake Superior and southern lakeshore cities in Illinois, Indiana, Michigan, and Ohio, soon to become the nation's industrial heartland. Copper and iron ore first moved in schooners, then in steamships designed for navigating the St. Mary's Falls Canal (now the Sault Ste. Marie or Soo Locks) at the eastern end of Lake Superior, quickly to become the world's busiest canal. Iron and steel hulls were better able to sustain the rigors of loading and transporting ore, as well as navigating the "steep" seas characteristic of the lakes. Because paddlewheels necessitated so much beam (width), it was on the lakes that propulsion by means of propellers—a variant of the ancient Archimedian screw—was first widely adopted. Propeller-driven ships became known simply as "propellers."[1]

At the turn of the 20th century, fully one-third of U.S. maritime commerce moved between ports on the Great Lakes. But the lakes are as much Canadian as American, and among Canadian merchants there were long-cherished dreams of the St. Lawrence River being opened to oceangoing ships and thus providing a direct gateway to Europe. The problem was a 240-foot difference in the water level of Lake Ontario and the Gulf of St. Lawrence. Before World War I, efforts to render difficult portions of the river more readily navigable—notably a stretch of rapids at La Chine, near Montreal—permitted the passage of 3,000-ton ships to and from the Atlantic. But this still was small by oceanic standards, and for decades U.S. railway and shipping interests opposed any further efforts to improve navigation, fearing that traffic would be diverted from eastern seaports.[2]

What broke the impasse was a changed political context, or, more precisely, Cold War geopolitical concerns about strategic isolation that seemed to outweigh parochial qualms. In 1954 the Canadian and U.S. governments struck a deal for a program to replace 22 small locks along the stretch from Ogdensburg, New York, to Montreal with seven big ones (Canadian contractors would build five of these) and thereby make the St. Lawrence into a "seaway" navigable by deep-draft vessels. The seaway opened in 1959 at a cost of $130 million to the United States and $260 million to Canada. Even though it would be seasonal, as the St. Lawrence is blocked by ice for at least four months (ever in quest of a "warm water" seaport, Canadians had financed a railway to Portland, Maine, in the 1840s), there was excited talk about a "new seacoast" for both nations.

Such visions reckoned without the advent of containerization and the hulking container ships that would dominate ocean commerce by the latter part of the 20th century. When the seaway opened, its 730-foot locks could accommodate more than three-fourths of the world's vessels. Today, three-fourths are too large. Hence, an oft-repeated story: a "lag" in infrastructure, and—in a political context much different from the 1950s—the likelihood of catching up seems nil. For one thing, everybody is sensitive to the potential for disabling a vast technological system with one small destructive device. And, even if billions to enlarge locks could be found in the public purse, there is no denying that the effects on the environment would be overwhelming, the opposition almost certainly intractable. Of course the environmental impact of the original project was enormous, too, but since then North American perspectives on such matters have been transformed. When it comes to technological change, ambient currents are always shifting.[3]

A half-century ago, the St. Lawrence Seaway was trumpeted as "the world's fourth greatest engineering project," and, in the tradition of projects like the Pennsylvania Main Line, it was surely a *technical* triumph: "To hell with the economics," said one enthusiast. "It is a magnificent conception, and it has got to be built." Even as international maritime commerce has increased 600 percent in thirty years, however, traffic on the seaway has actually declined. It may be too pessimistic to call it "an anachronism, frozen in time and likely to remain so," and yet there is no doubt about it belonging on a long tally of transportation projects that failed to work out as zealous proponents had anticipated.[4]

Which is not to say that the Great Lakes themselves are lacking maritime vitality. Even though only a small proportion of cargo goes through the seaway—mostly taconite pellets downbound from Newfoundland—and even though the volume of regional bulk transport remains about what it was when the seaway opened, commerce on the lakes exceeds by several times the volume that moves under the American flag on the high seas. Sinuous bulk carriers remain a familiar sight in the Soo's 1,200-foot locks, in the Welland Canal skirting Niagara Falls, and in lakeshore ports (Figure 14). Routes between such cities as Buffalo and Cleveland approximate the way a crow would fly, and for that reason even passenger liners kept operating after World War II. This was likewise the case with eastern riverways such as the Hudson, the avenues of trade and travel considered next.

6

RIVERWAYS

Today, among the sights we associate with major ports such as San Francisco and New York—where rivers broaden into harbors—are suspension bridges to facilitate travel and transport from one shore to another. With cables made of bamboo, such bridges date to the third century B.C.E. in China and with chains to the fourth century C.E. in India. But spans longer than a quarter-mile did not become common until after completion of the Brooklyn Bridge with cables spun from multiple strands of wire in 1883. Before such bridges (and before subaqueous tunnels), ports required transportation systems of their own, ferries. The first ferry in America began operation between Boston and Charlestown in 1631. It was powered by men wielding long oars called sweeps. After the Revolution, municipal authorities in New York and Philadelphia as well as Boston sought to implement service that was faster, and one response was the introduction of team ferries: ferry boats pulled by horses harnessed to capstans. At the same time, people began to experiment with systems that did away with animate power altogether, ferries powered by steam engines.[1]

Figure 14. Great Lakes "self-unloaders" such as the *Detroit Edison* were fitted with onboard conveyors that could disgorge a cargo of iron ore or coal in a few hours. Photograph by P. J. VanderLinden.

The Sociology of Invention

Besides harbors, steam power also seemed promising for the lower reaches of eastern watercourses, for Americans were well aware (as the expression went) that they "only crawled along the edge of the continent." Sailing ships plied the Connecticut, Delaware, Potomac, Susquehanna, and Hudson. But upstream voyages were awkward at best, even on the breadth of the Hudson where palisades deflected the wind. With steam power starting to appear in industrial settings, however, a door was opening and it did not require a leap of imagination to envision "harnessing" steam to a conveyance. Even though this idea occurred to a number of people at about the same time, Americans have always set store in designating the *one* inventor of any given device. Partly, in John Law's words, this is because of a tendency among those who chronicle invention to treat creative genius "as a process in which the end results of a system of collaborative production are gathered up and attributed to a particular individual." Such chronicles may also take up quarrels about priority with more passion than the contestants themselves. So it is that Robert Fulton got crowned as the inventor of the steamboat, when the process was actually quite complex.[2]

First, due notice needs to be taken of Europeans who succeeded in converting thermal energy into mechanical power as early as the 17th century. Credit must be shared even in the United States. Pride of precedence is due to John Fitch, who was born in Connecticut in 1744 but lived for many years near the Delaware River, working as a silversmith. In 1786, on the Philadelphia waterfront, Fitch and a mechanic, Henry Voight, began assembling a boat that used steam to power a set of duck-leg paddles (an adaptation reminiscent of Leonardo da Vinci's vision of an "ornithopter" with flapping wings for propulsion, in imitation of birds). In August 1787 they demonstrated it to delegates to the Constitutional Convention, seeking political favor in the form of a monopoly. The next year they completed another steamboat and in 1790 yet another with which Fitch inaugurated regular service to points upstream and down that lasted an entire summer.[3]

This was a substantial achievement, but Fitch also dissipated vast stores of emotional energy in contesting the claims of one James Rumsey. In 1784, Rumsey devised something he called a "stream boat" and showed it to George Washington at Berkeley Springs on the upper Potomac. Washington provided an affidavit crediting him with discovering "the art of propelling boats by mechanism." The "art" had nothing to do with steam power, and Rumsey did not get a steamboat in operation until after Fitch. But Rumsey encouraged the concept that priority was his, and the upshot was a wrenching dispute with Fitch, each accusing the other of fraud and deception. Even though Fitch secured monopolies on steam navigation in several mid-Atlantic states, as well as patents, his anger ultimately consumed him and he sank into a suicidal despondency.[4]

Fitch was of a type that fill the annals of technology. He could turn a promising idea into a functioning device, but he was not sufficiently sensitive to practical design, nor to the import of sustaining political favor. His eccentric behavior made it impossible to keep financial backers. "The day will come," he wrote, "when some powerful man will get fame and riches from my invention." This is just what happened. A generation of trial and error followed Fitch's experiments, notably by the Hoboken engineer John Stevens, who in 1804 built a small boat called *Little Juliana*, replete with a high-pressure engine linked to two shafts protruding through the stern and having "a number of arms with wings like those of a wind mill." We would call this twin-screw propulsion, and at the turn of the 19th century it was a remarkable novelty. Afterwards, Stephens and his son Robert built a sidewheeler, the *Phoenix*, which was steamed around Sandy Hook and 150 miles down the coast to be put in service on the Delaware River in 1809, thereby becoming the first American steamboat to navigate ocean waters, a decade before the *Savannah*.[5]

The *Phoenix* plied between Philadelphia and Trenton until 1814 when she was wrecked, a fairly long lifespan for an early-day steamboat. The reason she was taken to the Delaware in the first place was because there existed a monopoly on the Hudson granted by the state of New York. It was under this privilege that Robert Fulton, with his wealthy and influential backer (and father-in-law) Robert R. Livingston, had rendered the steamboat commercially viable. Scion of a political dynasty, Livingston's other claim to fame was the negotiation of the Louisiana Purchase from France in 1803—800,000 square-miles of territory stretching to the Rockies and more than doubling the size of the nation. Ultimately it was on the "western" river system, much of which flowed through this new territory, that the steamboat was given an indelible American character.[6]

As an expatriate artist, Fulton had begun his experiments with steam propulsion on the Seine in 1803, then continued them on the Hudson with Livingston's support. It was this support that got him through the door. In August 1807, a 133-foot vessel prosaically named *The Steamboat* made the 150 miles from New York to Albany in 32 hours (with an overnight time out at Livingston's estate, Clermont, the name later given to the boat). While Fitch and Voight had laboriously reinvented an older type of power plant, *The Steamboat* had a state-of-the-art engine imported from the premier English firm of Boulton and Watt. Sloops needed several days to tack their way to Albany. Now, traffic could move nearly as fast against the current as with it. By 1815, 20 steamboats had been built for enterprises in which Fulton was involved, one of which, the *Chancellor Livingston*, was capable of 8-1/2 miles an hour and could make Albany overnight.[7]

Fulton retained his monopoly until 1824, when a technological trajectory was diverted by a decision from Chief Justice John Marshall, a political man who stood by Federalist precepts about the Constitution even though the party itself was defunct. Fulton's steamboats plied both New York and New Jersey waters. In 1824, Marshall ruled that only Congress had jurisdiction over commerce involving more than one state, thereby invalidating Fulton's special privilege. A doorway was now open to others. In the commercial rivalry that ensued, powerful steamboats appeared on the Hudson and on Long Island Sound that were three times faster than the *Chancellor Livingston*. It was a story to be repeated again and again, actually two stories: many Americans would put a premium on "getting there first," and the deployment or redeployment of political power would often provide an inducement for significant technological change.[8]

After the advent of railroads in the 1830s, what continued to dominate steam navigation on the rivers of the eastern seaboard was the passenger trade, part of the story of *travel* in America; as long as steamboats booked fares on American rivers, the overnight run between New York and Albany remained the busiest. Yet steam played a much larger role beyond the Alleghenies, as its potential in commercial *transport* opened up the continent. In 1849, Stephen H. Long (who has appeared in the narrative before and will appear again) estimated that there were 17,000 miles of navigable riverway here.[9] There were more than that, if one really wanted to try. Ultimately some 3,000 steamboats worked waterways great and small—wherever, so it was said, there was enough water to take a bath (Figure 15). This required vessels of radically novel design.

Figure 15. As shown in this 1881 cut from *Harper's Weekly*, steamboats were then navigating the Red River of the North as far as Fargo, North Dakota, where the water was very shallow indeed.

Adaptation

The western rivers were vital avenues to the first settlers of European origin, just as they had been for Native Americans. Traffic was ordinarily one-way, rafts and flatboats getting dismantled at downstream destinations. More substantial keelboats could be worked back upstream by alternately poling, towing, rowing, and raising sail, but the process was just as arduous as it sounds; three months was the usual time it took to make the thousand miles from New Orleans to Louisville. Even the earliest steamboats could do this in less than a month and by the 1820s it took 10 days. In 1840 a French visitor remarked that the effects of steam navigation in the Mississippi Valley were "more striking than in any other part of the world." It was largely due to steamboats that, 20 years later, more than 15 million people lived west of the Alleghenies, nearly half the population of the United States. Technology per se may not cause change, but it surely can enable people to change a way of life.[10]

Steam had made its debut on the Ohio in 1811, when the *New Orleans* was completed at Pittsburgh on the basis of plans supplied by Fulton, who coveted a monopoly on western commerce such as he then held at home. But westerners did not take kindly to would-be monopolists from back east, and Fulton was perceived not as a benefactor but rather as a "grasping mercenary." It also turned out that designs suited to the Hudson were not well adapted to western riverways, often shallower and swifter. In the late summer, operation had to be confined to the placid 300-mile stretch of the Mississippi between Natchez and New Orleans. A Philadelphia mechanic, Oliver Evans, first perceived the need for hulls that did not sit deeply in the water and for engines more powerful than the low-pressure Boulton and Watt type. The westerner who most quickly recognized both necessities was Henry M. Shreve, who had been taking keelboats down the Ohio since 1799. They found expression in Shreve's *Washington*—much the largest steamboat in the west but still able to navigate the Ohio up to Pittsburgh and even further. By placing the boiler on deck instead of in the hold, Shreve anticipated the eventual deck-mounting of engines as well, thereby permitting very shallow draft. Initally, in Louis Hunter's words, the western steamboat was "a poorly coordinated combination of a seagoing vessel and a stationary engine, neither of which was suited to the use to which it was put." But it was soon refined, by men with little formal training, into a marvelously functional conveyance—albeit one that wasted fuel, wore out quickly, and, worst, was prone to explode.[11]

POLITICS, TECHNOLOGY, AND DANGER

As currently attested by more than 40,000 highway fatalities every year, technologies we develop often yield results we do not want along with what we do want. Fulton's engines used the expansive force of steam at a modest pressure of seven pounds-per-square-inch (PSI) above the pressure of the atmosphere. Even before 1807, however, Oliver Evans—and, independently, Richard Trevithick in England—had devised a type of engine that operated at 30 PSI and had the advantage, insofar as conveyances were concerned, of being far more compact. Knowing the destructive potential of high-pressure steam, Evans carefully specified rules for safe boiler design. When these went unheeded, disaster was inevitable. By 1817, there had been 25 fatalities on the Ohio and Mississippi due to boiler explosions, including 13 killed when Shreve's *Washington* exploded at Marietta, Ohio.[12]

The toll would grow worse, much worse, but explosions were not the only hazard. Running aground was common on the Missouri and Yellowstone, rivers flowing easterly off the plains, even though many vessels had the paddlewheel mounted aft in order to thread a narrow channel (Figure 16), much as Great Lakes steamers would be fitted with propellers in order to navigate locks. Grounding on a sandbar might just be inconvenient, but far less forgiving were snags, trees that had fallen into the water after their roots were eroded and then held fast to the bottom. Sometimes these were called sawyers because, like an iceberg, they could saw a hull wide open. In the 1820s, Shreve—having won government appointment as superintendent of western river improvements—developed a floating device similar to the one used to extract stumps along the route of the Erie Canal (Figure 17). His "snag puller" proved its worth with the notorious Red River Raft, a chaos of logs that blocked navigation above Nachitoches and had backed up for *160 miles* by the time Shreve attacked it. It took five years to get rid of the obstructions, but he finally opened the river to a town in the northwestern corner of Louisiana, now designated a "port" and called Shreveport.[13]

Figure 16. This drawing shows how a Missouri River sternwheeler could navigate shoal waters. It was not uncommon for such vessels to float 200 tons of cargo through narrow channels only 36 inches deep, and paddlewheels located astern were less vulnerable to damage. In the event of grounding, the "grasshoppers" at the bow could be wielded like crutches to "walk" into deeper water. Intended for the same purpose, though never adopted, were the "adjustable buoyant chambers" patented by Illinois Congressman Abraham Lincoln in 1849.

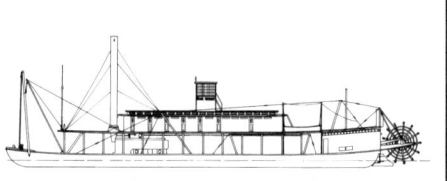

Thanks to political logrolling and the labors of the Army Corps of Engineers, diesel towboats (a misnomer—actually they push strings of barges) presently make their way hazard-free to cities that might seem even less likely to have ports, such as Tulsa, on the Arkansas River. The work performed by the corps has amounted to vital support for the towboating industry. It has also remained controversial, especially projects that critics regard as pure "pork" such as the Arkansas River Navigation and the "Tenn-Tom" connecting the Tennessee River in Mississippi with the Tombigbee in Alabama.

Even more contentious at one time—strange as it seems—were efforts to prevent boiler explosions. Commercial and manufacturing interests had routinely

Figure 17. The sidewheeler *Missouri* was an early 20th-century snag-puller that closely reflected Henry Shreve's initial 1830s design. It is seen at St. Louis against a backdrop of the Eads Bridge. Courtesy Missouri Historical Society.

sought a helping hand from the federal government in the form of patents, land grants, and tariffs. But demands made on behalf of the "general welfare" were unprecedented, and it was steamboat entrepreneurs who first felt such demands. No matter how insistently they cried that inspection was unconstitutional, even despotic, there was no disputing that steamboat boilers could be lethal. Sometimes the cause was poor workmanship, materials, or maintenance; sometimes the men who operated boats were directly to blame, as when they tied down "safety" valves, designed to vent excess pressure, in order to win a race.

Although there had already been many fatalities, in 1838 the roof fell in. First, the *Oronoko* blew up near Natchez, killing more than 100; then, a few days later, the *Moselle* blew up at Cincinnati, killing 150. At that point, the government stepped in, but very tentatively, and annual death tolls actually mounted rather than diminishing; in a period of less than three years after 1850, nearly 700 perished as a result of boiler explosions. After that, Congress took decisive action, a 44-point law with enforcement assigned to the Treasury Department. The Steamboat-Inspection Act of 1852 subjected not only the boiler but also the hull and emergency gear of every steamboat to rigorous scrutiny and also required licensing of engineers, masters, mates, and pilots.[14]

Resistance to the perceived abuse of governmental power was (and is) an article of faith among many Americans, and even a roving Treasury Department agent deemed the power to refuse or revoke a license—and thus deprive somebody of a means of making a living—"monstrous." But, he added, it was intended to avert "a monstrous evil." Resentment died slowly. There was no arguing with the numbers, however, 35 percent fewer deaths from steamboat explosions in the eight years after 1852 than in the eight years before. And yet the most appalling of all disasters was still to come. In April 1865 the *Sultana* blew up at Memphis, killing 1,700, mostly Union Army prisoners of war who had just been released. What was notable about the *Sultana* was her experimental

Figure 18. Pausing from work of a project to improve navigation on the Cumberland River above Nashville in 1891, black laborers pose for a photographer. *Improve* is of course a loaded word, and any such project would be freighted with unanticipated consequences. Courtesy Prints and Photographs Division, Library of Congress.

boilers designed to minimize weight and fuel consumption, devices with which inspectors may have been unfamiliar, an invitation that designing engineers probably should have declined.[15]

POLITICAL SUSTENANCE

By the late 1860s riverboat travel was under sustained assault from the men who operated railways, who enjoyed several distinct competitive advantages. One was relative freedom from seasonal vagaries, low water and ice. Another was greater speed and yet another was more direct routes. It took 16 hours to travel from Cincinnati to St. Louis by train, three days by boat. On the lower Ohio and the Mississippi, passenger traffic held up into the 1870s, as operators offered accommodations aboard stylish Gothic sidewheelers. Still, as the operators of luxury liners and "extra fare" streamlined trains were to discover in the 20th century, when it became possible to get there quickly, most travelers preferred that to getting there in style, and so passenger traffic suffered the same fate at the hands of the railways that the railways would later suffer at the hand of the airlines.[16]

Nevertheless, the riverways still flourish on the basis of bulk cargo. The last generation of sternwheelers was designed simply to push barges and eventually these gave way to the powerful towboats that now ply an empire whose main artery is the Mississippi. It is a remarkably efficient technology, but one needs to bear in mind that the cost of 1,200-foot locks and other elaborate infrastructure is distributed among the taxpaying public, and that this technology thrives as it does because of sustenance from political precincts.

When steamboats first navigated the Ohio, during seasonal stages of high water it looked like a serpentine 1,000-mile lake extending from Pittsburgh to Cairo, Illinois. But slack water exposed hundreds of sandbars and stone ledges, and at Louisville the river fell 22 feet through whitewater rapids. This would have been an insuperable impediment had not work begun on "canalizing" the natural watercourse in the late 1820s. Since the Civil War, the Corps of Engineers has rendered some 26,000 miles of waterways navigable to vessels drawing nine feet (Figure 18). Besides Tulsa, tows of barges presently reach Minneapolis, Chicago, Sioux City in Iowa, Birmingham in Alabama, even western Virginia. The Tenn-Tom, providing a second outlet to the Gulf of Mexico from the Ohio, stands as the nation's costliest waterway. Like the St. Lawrence Seaway, however, it is limited by the size of its locks and there is not much traffic. The Tenn-Tom, writes its historian, Jeffrey Stine, "found strong support among regional boosters, chambers of commerce, investors, bankers, and commercial entrepreneurs"—the expenditure was not theirs—but clearly he agrees with the editorialist who called it "the world's largest wet elephant."[17]

Pork, elephants, whatever, such projects have received faithful support from Washington, and, even though the corps is typically embattled, barge lines often prosper. Among American transportation systems there has nearly always been this link between politics and prosperity, and never more so than with the railways. To the railways, a hand was first extended and then it was withdrawn; Edward K. Collins would have felt the pain.

7

Railways

A 21st-century diesel towboat, behind a phalanx of 195-foot barges lashed together with stainless-steel cables, may be pushing 15,000 tons of cargo, perhaps limestone, gravel, or cement, perhaps steel, grain, or petroleum. But that still does not match the tonnage of a railway train headed for eastern power plants with coal from Wyoming's Powder River Basin. Almost from the beginning, capacity was where railways had a competitive edge. At first, canal boats were drawn by a horse or mule trudging along a towpath, and typically had a capacity of 25 tons, compared to 10 tons for boxcars. One beast of burden, one boy ashore, one boat with another boy aboard—this was a combination that fit only with fading agrarian ideals. By the end of the century, canal boats often had a 200-ton capacity, eight times the capacity of a boxcar. But of course railway cars move in *trains* and the tonnage of an average train was approaching 300.[1]

A railway had advantages deriving from circumstances that were not socially malleable. At a walk, a horse could pull a canal boat with a 25-ton load, more than three times as much as on a railway. At a trot, however, a horse could barely handle a canal boat at all, due to resistance of the water, but it could pull almost as much on a railway as when at a walk. Moreover, railways were seldom fazed by inclement weather, while a season on many waterways was (and is) only eight months. Railway routes did not always describe a beeline from point to point, but few rivers came close. On the Ohio, the distance from Pittsburgh to Cincinnati was 470 miles; via the railway completed in 1853, it was 316.[2]

One could find descriptions of conveyances running on flanged wheels in Georgius Agricola's *De Re Metallica* 300 years earlier, and the virtues of a system in which wheels were guided by rails were fully understood by the 18th century. After steamboats appeared on America's riverways, a door was open for the steam railway. Among its many heralds was Benjamin Dearborn. One day in 1819, Senator Harrison Gray Otis of Massachusetts read a memorial from his constituent, Dearborn, who was enthusiastic about the concept of "carriages propelled by steam on level rail-roads." Dearborn acknowledged the obvious, that railways were nothing new, nor were steam carriages: in Hoboken, the prolific John Stevens had already made such a conveyance, and, before that, both Oliver Evans and Richard Trevithick. And even before that, Nicholas Cugnot, a Swiss engineer, had devised a steam tractor for the French army. Perhaps neither Dearborn nor Otis knew about Cugnot, though they were quite familiar with

Evans, and also with the railway that Gridley Bryant had built in 1805 to move earth in conjunction with residential development on the north side of Beacon Hill. Perhaps Otis and Dearborn were under the impression that nobody had thought in terms of steam carriages with flanged wheels running on rails, but Trevithick had actually secured an English patent for just such a combination.[3]

All of which is by way of indicating that, like the steamboat, the railway had a "prehistory" prior to 1825, when the first common-carrier—that is, a venture soliciting patronage from the general public—with locomotive power began operation. This was George Stephenson's Stockton and Darlington in Durham, England.[4]

Technological Transfer

New Yorkers always had a keen appreciation for transportation innovations, and in fact the first locomotive built in the United States came from the West Point Foundry in New York City. But New Yorkers had just celebrated the opening of the Erie Canal, tapping a western hinterland. Thus, enthusiasm for railways was strongest in cities which had not yet done so with their own hinterland, including Boston, Charleston, and particularly Baltimore, the third largest city in the country with a population nearing 80,000. For none of these cities did a canal like the Erie appear to be a viable option. A railroad was.[5]

Though "firsts" are vexed, as we have seen, there is general agreement about the first common-carrier railway in North America. In 1827, a group of Baltimore merchants chartered the Baltimore & Ohio Railroad and on July 4, 1828, the first of its stone ties was set in place, the guest of honor being Charles Carroll, the sole surviving signer of the Declaration of Independence. By May 1830 the B&O had completed 13 miles of track running westerly to Ellicott's Mills—but, no, this did not fulfill Ben Dearborn's vision because the B&O initially used horsepower, as would a canal. Rather, the first "carriages propelled by steam" ran on the rails of the South Carolina Canal and Railroad Company, which owned that locomotive from New York, *The Best Friend of Charleston*. It was one of the few times that southerners would step through an open doorway first. What they expected Charleston's best friend to do was divert cotton traffic from arch-rival Savannah, and by 1833 this line had 136 miles of track, running into the bottomlands around Hamburg. It was the longest continuous stretch of railway in the world.[6]

By 1833 there were several other lines in operation, including the Boston & Worcester and the Mohawk & Hudson from Albany to Schenectady (initially, there was a statutory ban on competing with the Erie Canal for freight traffic). Within four more years, the B&O had reached Harpers Ferry as well as extending tracks southward to Washington, D.C., and there were lines running from Boston to Lowell and Providence (and by 1841 to a connection via Worcester with the Erie Canal at Albany). John Stevens's son, Robert, whom we have already encountered in maritime contexts, had completed a line for the Camden & Amboy Railroad and Transportation Company between lower New York Harbor and the Delaware River, a premier stagecoach route for more than a century.[7]

It bears repeating that no technology is ever introduced without some people losing as others gain. Railways were no boon to coachmen or innkeepers, and certainly not to canal companies (unless one firm monopolized both forms of transport, as did happen). But consider somebody who traveled on business. In 1830, such a traveler could have moved no faster than a galloping horse might carry him; within three years, the Camden & Amboy's *John Bull*—delivered from the Liverpool manufactory of Robert Stevenson, George's son—was steaming at 35 miles an hour across the New Jersey flatlands.[8]

As already evident with riverboats and soon to be reemphasized with clipper ships and oceanic sidewheelers, many Americans considered speed a valuable commodity. "Railway mania" matched and then overwhelmed canal mania. By 1840, 60 different companies were operating 3,228 miles of line, almost twice the mileage throughout Europe. A decade later, there were 5,600 miles of railway in New York and Pennsylvania alone, including the Pennsylvania Railroad's nearly-completed link between Philadelphia and Pittsburgh; soon, Erastus Corning would merge a series of short-lines strung out across the Mohawk Valley from Albany to Buffalo into the New York Central.[9]

Railways connecting major cities would become known as *trunks*, as distinct from branches. At mid-century, trunks and branches together represented an investment of $300 million. The cost-per-mile of infrastructure averaged $24,000 in the south, nearly twice that in the mid-Atlantic states, largely because labor was paid rather than forced. Every single state had some trackage, although there were only 2,100 miles in all nine southern states. Not all mileage, north or south, was profitable, but, as with canals, railways emerged "in a climate where hope and imagination could matter more than cold economic calculus." There had been some publicly funded construction on the order of the Erie Canal, and New York, Massachusetts, Maryland, and Virginia had provided cash or credit to several railways. The federal government had financed various surveys and even before the Civil War made land grants of nearly 25 million acres. In overwhelming measure, however, railways had been financed by means of capital stock and bonds (in about equal proportions), and so far hardly any of the funds had come from abroad.[10]

Adaptation

It was different with technology. Both the steam locomotive and the railway itself had English origins, and English locomotives were imported until 1841, 120 of them all told. With that in mind, historians have typically constructed a narrative in which English technology diffused to America and the rest of the world. Yet American locomotives were being exported as early as 1836 (to Russia) and some argue for two distinct traditions developing more or less simultaneously. In contrast to the British Isles, the prosperity of railways in the United States was often contingent on the development of new markets, not tapping markets that already existed. Lines had to be built on the cheap. This meant spending as little as possible on grading, bridges, and tunnels to minimize grades and curvatures. From the start, what compensated for steep climbs and sharp curves was sheer power. Indeed, North American railways came to resemble a mirror image of those in Britain, with their meticulous infrastructure and anemic locomotives.[11]

Almost all aspects of locomotive design had been modified by 1850. While America's canals embodied scant technological novelty, its railways and locomotives—as with its flat-bottomed sternwheelers—became distinctive. Initially, tight curves had caused frequent derailments of locomotives with four driving wheels and two axles rigidly anchored to the frame, standard British design. In response, John B. Jervis of the Mohawk & Hudson devised a four-wheeled "truck" that replaced the front set of drivers and provided guidance. Then, in 1836, Henry Campbell of the Philadelphia, Germantown & Norristown patented a locomotive with a truck and four drivers as well, which became known as the "American Standard" and would predominate on U.S. railways until the 1890s and even longer in "express" passenger service. Other innovations followed: locomotives with six and eight driving wheels (and eventually as many as sixteen), anthracite-burning fireboxes, trailing trucks, enclosed cabs, mechanical stokers, and many more; later, American railways pioneered in outfitting cars with air-brakes and automatic couplers (which freed brakemen from their deadly dance between moving cars), and in devising specialized cars for livestock, petroleum, and refrigerated transport.[12] (See Figure 19.)

Figure 19. One of a fleet of cars built in 1898 by the Pullman Co. and outfitted with ice bunkers for shipping apples and peaches. When the bankrupt Colorado Midland was abandoned in 1917, much of the equipment went to France for the American Expeditionary Force, but not the refrigerator cars. Courtesy Kalmbach Publishing Co.

British railways were fenced, but investors in the United States tried to spare this expense, often to dire consequence. Palliatives included locomotives with bells and whistles as warning devices, oil-burning headlights, and finally the ubiquitous "cowcatcher," which, in theory at least, would spare livestock (or pedestrians) that wandered onto the track. In railway shops, a new class of master mechanics reigned supreme, men who had little use for abstract knowledge, loved novelty, and resisted standardization. There were no locomotives anywhere else in the world as powerful as American locomotives. This was true whether they were built in railway shops or commercially. As early as 1832, Matthias Baldwin and William Norris were both producing locomotives in Philadelphia, the latter with Stephen H. Long devising novel but impractical components, "a victim of his own imagination." Baldwin, on the other hand, simply copied Jervis's tried-and-true designs at first and by the 1840s was on his way to becoming the fourth-largest American industrial employer; at its peak in the early 20th century the Baldwin Locomotive Works would employ 18,500 workers and outshop as many as 3,500 locomotives in a single year.[13]

As for infrastructure, stone blocks under the track, or wood pilings (as used in South Carolina), soon gave way to wooden "ties." Wooden rails surfaced with strap iron were superseded by iron rails with a cross-section shaped like an **I**, though called **T**-rail—yet another novelty introduced by Robert Stevens—which were attached to ties with spikes that hooked over the base. Because the possibility of interconnecting lines was rarely considered at the outset, the gauge (distance between the rails) varied: John Stover writes that "Teamsters, porters, and tavern keepers were happy that not a single rail line entering either Philadelphia or Richmond made a direct connection with any other railroad entering the city." While lines in most of New England and parts of the mid-Atlantic emulated the gauge of Stevenson's railways in England, 4 feet 8-1/2 inches, some were as wide as six feet and many in the south were five. Even after most lines were standardized so that freight cars could be interchanged among different railways, in Maine a few two-foot-gauge "baby railroads" operated through the 1930s and some lines in the rugged confines of the Rockies stayed with three feet.[14] (See Figure 20.)

Freight cars initially had four wheels rigidly attached to the frame in accord with English practice, but this was soon superseded by four-wheeled swivel trucks at either end, giving cars much greater capacity. There was nothing in England or Europe like American passenger cars. At first these resembled stagecoaches, and they were still called coaches in the 1880s, when they were often 50 feet long with accommodations for 60 passengers; many had toilets and drinking water, a few even had sleeping accommodations. For two decades, passenger revenue exceeded freight, but by the 1850s they were about equal and by 1880 freight traffic accounted for nearly two-thirds of receipts.[15]

Figure 20. The largest and last operator of narrow-gauge lines in the Rockies was the Denver & Rio Grande. Three locomotives—two pulling, one pushing—are seen here working a short train up a heavy grade. Fred Jukes photograph, courtesy Railway & Locomotive Historical Society.

TECHNOLOGY, POLITICS, AND WAR

When Abraham Lincoln was elected president in 1860 and disunion loomed, there were 30,000 miles of railway in the United States, the total investment well over a billion dollars. With 5 percent of the population, the United States had nearly as much railway mileage as the rest of the world combined. As business operations, there was nothing even close to the American railways in scale and complexity. Railways gave an enormous impetus to other industries such as coal-mining and lumbering, not to mention telegraphy, and after the war the railways would essentially create the steel industry. That railways were indispensable to American economic growth was an all-but-universal perception.[16]

In the iconoclastic academic climate of the 1960s, however, an economic historian who later got a Nobel Prize sought to refute this. By the end of the 19th century, said Robert Fogel, railways had contributed less than 5 percent of gross national product. Had they never existed, more investment would have flowed into canals (which could have been heated for year-round operation) and motorized highway vehicles would probably have appeared sooner. It was an intriguing argument and Fogel scored notable points—for example, in pointing out that the British Industrial Revolution "was virtually completed before the first railroad was built."[17]

Fogel based his hypothesis on the proposition that one could substitute freely among "exogenous" technologies; technologies were nothing more than *black boxes*. Even if this were true, Fogel did not seem to have thought very much about doorways that might or might not be open, and rarely has an economic analysis been set forth that was so oblivious to the ways in which people construct a machine's social reality "as part of many human situations which collectively define its meaning," as David Nye put it. To the American public, the railway—and particularly the railway locomotive—was anything but an inert black box. No other device came close to the railway locomotive in its capacity to stir human emotions, to inspire visions of empire or apprehensions of calamity. "People plan and try to execute rational strategies for promoting or resisting a given technology," writes John Staudenmaier, "but these same people also respond to technology affectively, *with awe or fear or anger or enthusiasm*. We need to learn to understand our technological behavior as a constant blend of these very different modes of consciousness."[18]

Nowhere was this borne out more clearly than with the multiple intersections of technological change and the impending crisis. Even though the Erie Canal tapped markets in the vicinity of Lake Ontario and Lake Erie, much of the commerce of the Old Northwest (the five states bordering on Lake Michigan and the western part of Lake Erie) had been going southward to New Orleans. But after mid-century Chicago became the hub of 11 different railways, nearly all of which looked east or west, not south. Chicago, St. Louis, and Cincinnati were tied directly to New York, Philadelphia, and Baltimore. Completion of these trunks resulted in a realignment of commercial allegiances, with more and more trade heading for the eastern seaboard rather than down the Mississippi. Railway rates were higher, but in terms of speed there was no comparison, nor was there any comparison with the railway's capacity to tap local and regional trading centers via branch lines.[19]

And even as traffic flows shifted from north–south to west–east, even as new sectional economic alliances were being forged, a political party was arising whose guiding purpose—opposition to the extension of slavery into the territories—was anathema to southern power-brokers. Railways epitomized the reality that these men could no longer think in terms of shared politics with people in Illinois, Indiana, or Ohio. What was commercial was also political. Railways seemed to unify the populace in the north, but divide rich man from poor man in the south; both tangibly and symbolically, railways played a central role in the coming of the Civil War.[20]

They played an even more central role in the war itself. In 1861, the states of the Confederacy had only about a third of the nation's railway mileage. The infrastructure was weak and everything from locomotives to lubricating oil came from northern suppliers. Many of the men involved in engineering and operations were northerners who left after the outbreak of war, or, if they remained, were regarded as untrustworthy. During the war, much of the southern railway system was ruined, but in the north only the Baltimore & Ohio suffered serious damage.[21]

The Transcontinental

Then, there were the collateral effects of disunion. Although the firm of Russell, Majors, and Waddell served supply routes in the west with Conestoga wagons, and Ben Holliday operated hundreds of Concord stages, the prospect of a transcontinental railway had long stirred popular passions. Army engineers had surveyed various potential routes, and the least costly to build seemed to be a southern route along the 35th parallel. But any such project was dead from the instant of secession. On July 1, 1862, President Lincoln signed the Pacific Railroad Act, authorizing the Union Pacific to build westward from Nebraska and, somewhere in Utah, to meet up with the Central Pacific, building eastward from California. This legislation also authorized land grants of 12,800 acres per mile—land with which the railways hoped to lure settlers onto empty plains and thus garner traffic—and it provided for government loans whose amount ranged according to the difficulty of terrain. This was one factor contributing to the episodes of corruption that accompanied construction of the Pacific Railroad. Most notorious was the Crédit Mobilier, a dummy firm designed to provide kickbacks to Oakes Ames, whose company had been supplying tools for large-scale construction projects all the way back to the Middlesex Canal.[22]

Eventually there would be several transcontinental railway lines, even southern routes. But the completion of the first, on May 10, 1869, at Promontory, on the north side of Great Salt Lake, was an event of transcendent import. First and foremost, it established a precedent for political support of big business that was masked in free-market rhetoric (socialism for the rich, some would call it). It also loosed fears of corporations so ruthless that they posed an immediate threat to democratic values, fears reinforced by fire-and-brimstone locomotives that were in stark contrast to the bucolic imagery of the canal boat, or even western steamboats, "light and fragile as pasteboard."[23] Canal and steamboat companies had been managed with methods not unlike those employed in Renaissance Venice; railways devised new systems of accounting that seemed impossibly opaque to outsiders. As to finance: chicanery, secrecy, and overvaluation seemed to be the corporate norm. (And none of this will sound unfamiliar at the beginning of the 21st century.)

In operations, managers pursued strategies they regarded as self-evidently rational, later "scientific," one notable instance of this involving the creation of standard time-zones known as railway time. In the face of demands for unflinching discipline, workforces became increasingly restive; strikes were brutally broken, sometimes with federal troops, while lavish dividends continued to flow. The remark of one railway baron that "the public be damned" echoed and reechoed. Even as shippers were confronting crushing rates and soldiers were shooting strikers, the railway network kept growing at an accelerated pace—an average of 7,000 miles of new track every year throughout the 1880s, when land-grants became a prime issue in national politics.[24] (See Figure 21.)

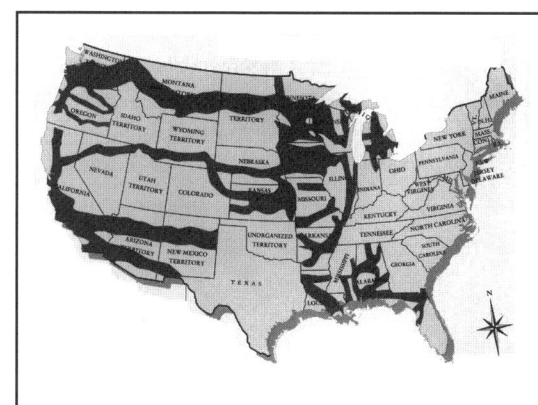

Figure 21. This map, published during the 1884 presidential campaign, was titled "How the Public Domain has been Squandered by Republican Congresses." It exaggerated the actual extent of land grants by about 400 percent, but, even so, got copied in American schoolbooks for several generations.

POLITICS, TECHNOLOGY, AND ENTERPRISE

Attempts by individual states to regulate collusive rate-making and kickbacks were futile. The Interstate Commerce Commission (ICC), established in 1887, was as ineffective as the initial response to steamboat explosions, serving largely as a tool of the interests it was supposed to be regulating. What finally did put the brakes on the railways, however, was a severe economic downturn beginning in 1893, which sent much of the nation's mileage into receivership and soon put financial control in the hands of Wall Street. This set the stage for an ironic turn of events. Fearing the specter of government ownership—a campaign cry raised by William Jennings Bryan—Teddy Roosevelt bore down heavily on the "railroad problem" after he became president. The Hepburn Act of 1906 gave the ICC unilateral power to determine "just and reasonable" freight rates, and its power was even further magnified in 1910 and again in 1913.

And therein lay the irony. As the ICC's coercive authority was being reinforced, the railways themselves were on the verge of a general economic decline. Inflation, along with the advent of highway competition, was choking off profits even as regulatory measures impeded the capacity of many firms to raise capital for modernization; Albro Martin has called this situation "enterprise denied." Nearly all main-line track had been replaced with steel by the turn of the century, a massive undertaking, but only one facet of the needed improvements. New locomotives, new steel freight cars and coaches, new terminals—these were within the reach only of wealthy railways like the Pennsylvania and the New York Central (NYC). The Pennsylvania would tunnel under the Hudson River to reach its new Manhattan station, the NYC would build Grand Central, the Pennsylvania and a few others would electrify segments of main line, the Burlington and Union

Pacific would inaugurate lightweight streamliners. Meanwhile, the New York, New Haven & Hartford, the Chicago & North Western, the Erie, and many others would slide into bankruptcy.[25]

Operating innovations would include Centralized Traffic Control, with all signals and switches tied to a single command center. Railways would enhance the performance of locomotives with devices to heat feedwater and "superheat" steam in the cylinders. Western lines in particular would favor articulated engines, some weighing a million pounds, reminiscent of the giant schooners of a generation past. And then, when those engines were only a few years old, these lines would join in an incredibly swift industry-wide switch to diesels. Railways were still ordering new steam engines in the late 1940s; steam was gone within a decade, even from lines that banked on coal traffic. Beyond loans from the Reconstruction Finance Corp., funds for "dieselization" were not always easy to find, but to have tried to sustain fleets of coal-burning locomotives would have been suicidal, particularly with John L. Lewis and his miners holding railways virtually hostage.[26]

After deregulation in the 1980s, there would be about 28,000 diesel locomotives in service nationwide and 1.7 million freight cars. Most were fairly new, but the industry as a whole would have endured decades of hard times. Total mileage had shrunk by one-third as lonely branches were abandoned and major carriers went out of business as well—the Rock Island Line, for one, and even a transcontinental, the Milwaukee Road. A botched attempt to merge the Pennsylvania and the New York Central in the late 1960s resulted in the largest business failure in history.[27]

In many respects the railways were never the same as they had been on the eve of World War I, when the extent of the network peaked at 260,000 miles of track, 25 percent of the world's total. When the United States declared war in March 1917, the president of the B&O, Daniel Willard, foresaw unprecedented demands on capacity and sought to institute a plan to operate all lines together as an integrated "continental system." But that would have violated antitrust laws. As it turned out, within a few months hundreds of thousands of freight cars headed for the East Coast were stalled in classification yards, and even when shipments arrived at ports of debarkation there was no place to unload. The system had failed. On December 17, President Woodrow Wilson named his treasury secretary, William G. McAdoo, "director general of the railroads." The most powerful industrial enterprise in the world, unchallenged and unchallengeable less than a generation earlier, was under the control of the United States Railroad Administration. In a word that had struck terror into the hearts of both Main Street and Wall Street ever since Bryan used it while campaigning at the turn of the century, the railways had been nationalized.[28]

8

HIGHWAYS

Between 1880 and 1900, 100,000 miles of new railway were completed. Between 1880 and 1910, the number of freight cars in service quadrupled, from 539,200 to 2.2 million. In 1910, the railways booked nearly a billion passengers. Railways dominated not just travel and transport, but the entire American economy (Figure 22). Railway investment was larger than the investment in the entire manufacturing sector. True, the absence of all-weather roads linking farms and railway depots had long been an irritant to rural America, a lack decried by the Populists and other partisans of the farmer. Yet hardly anyone envisioned roads as an *alternative* to rails; at most, roads and the trucks using them would simply expand the territory from which railways could draw shipments. The catchword for improved highways was "feeders."[1]

When money for the National Road had run out in the 1830s, maintenance was turned over to the states, which seldom took serious responsibility. By the time of the Civil War, surfaced roads rarely extended beyond city limits. What began to change this situation was people stepping through a doorway and mounting a new "safety" bicycle—chain driven, with pneumatic tires and both wheels the same size, which decreased the risk of getting pitched over the handlebars. By the 1890s, hundreds of firms were making bicycles by the millions. The League of American Wheelmen was founded with the aim of fostering improved roads for recreational cyclists. In response, Congress established the Office of Road Inquiry (ORI) in 1893. The ORI's sights were set on a system of "good roads,"

Figure 22. Encouraging tourism in national parks was a central part of the marketing strategy for western railways. Typical was the Santa Fe's promotion of the Grand Canyon.

not only for the convenience of bicyclists, but also for improving the accessibility of markets and bringing to farmers the social and material resources of towns and cities.[2]

At the turn of the century, perhaps betraying a managerial class growing myopic, railways actually *supported* the good roads movement. Several of them operated special trains aimed at fostering enthusiasm. This support held up even as evidence began to indicate that some short-haul freight traffic was going via highway rather than railway, and even as Congress got past any constitutional scruples and began appropriating funds for highways, which it first did in 1916. It was still hard to imagine trucks as anything more than "supplementary" to railways. Then, the system broke down. Railways proved incapable of handling the flood of war materiel destined for France or even delivering the coal essential to heating homes on the eastern seaboard.[3]

TRUCKS FOR THE LONG HAUL

Comparative studies had indicated that a five-ton truck was twice as expensive to operate as a wagon with a three-horse team, given the same load, but that the truck's greater speed resulted in less than half the ton-mile cost of the team and wagon. This lesson was not lost on anyone responsible for large numbers of wagons and teams. Even before the United States entered the war in 1917, more than 40,000 American trucks and ambulances had been delivered to the Allies in Europe. By the Armistice, three times that many were being operated by the American armed forces. A member of the British War Cabinet remarked that "the war could not have been won if it had not been for the great fleets of motor trucks." The Allied cause "floated to victory on a wave of oil."[4]

In the United States, there were more than half a million trucks in operation by 1919, many of them built to rugged army specifications. During the interwar years, with railways—now back in private hands—still seeking "coordination" of all forms of transport and still treating trucks as short-haul auxiliaries, over-the-road *long-haul* trucking began to take off (Figure 23). As Congress voted generous appropriations for highway improvements during the 1930s, intercity truck traffic more than doubled even as railway management remained hobbled by the mandates of the Interstate

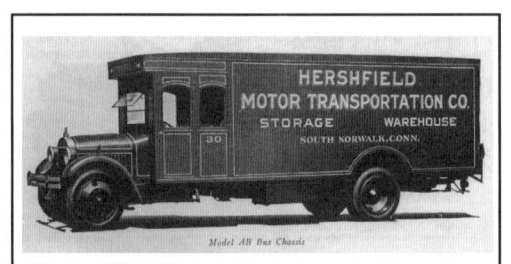

Figure 23. Door-to-door household moving was one of the first niche markets for long-haul truckers. When the Mack Truck Co. began manufacturing these vans in the 1920s, it promoted them on the basis of their speed, capacity, reliability, and, something new, "advertising value."

Commerce Commission. While a 1935 act did impose regulations on motor carriers, these were never as onerous as the ICC's. Even with bankruptcy dogging many railways, the ICC acted as if it were still the days of robber barons, before the Hepburn Act and begun to impede vital avenues of investment. But there was plenty of blame to go around. A good part was due to railway management that was excessively cautious about innovation, about venturing through open doors. For example, it seemed clear that a reciprocating steam engine was actually quite an inefficient device for converting heat into energy. A diesel engine, in which the fuel-air mixture is ignited by compression, has a much higher thermal efficiency, and ships were being outfitted with diesel powerplants by the 1920s. At the same time, as Steven Usselman writes, "railroading had slipped into a long, steady decline during which it acquired the trappings of a technical backwater." Certainly it is telling that when railways finally adopted the new form of motive power, the primary supplier was not one of the old-line locomotive firms like Baldwin, it was General Motors.[5]

Most trucking companies started out as family businesses and some of the largest remained in control of hard-driving entrepreneurs of the sort who once dominated the railways. During the second half of the 20th century, an innovative push came from a firm called Sea-Land, containerization, which enabled multi-modal shipment, by rail, ship, or truck without any need to "break bulk." Rudi Volti writes that in the era of globalization, containers have been "a key element in an international economic system that puts Korean television sets in American living rooms and German machine tools in Brazilian factories." At the beginning of the 21st century most of the historic railway logos like the Pennsylvania's keystone represent "fallen flags," with just a handful of huge systems dominating the industry. They thrive on containerization, and yet railways carry only 15.8 percent of intercity freight by weight and 5.7 percent by value. In terms of gross revenue, trucks account for 86.5 percent.[6]

It is important to note that neither form of transport is exclusively a "guy thing" anymore, as was the case with every mode so far addressed in this booklet: traditionally, an oceangoing freighter was "no place for a woman," nor was the conductor's caboose of a railway train or the cab of a Mack truck. By 1999, Linda Nieman was a Union Pacific conductor with top seniority, but it was different in 1979, when she hired on as a brakeman at age 33. Wherever she went she was told that for a girl the work would be impossibly difficult. Women now command 18-wheelers as well as freight trains, but, for women who sought such jobs a generation ago, penetrating male space was no easy matter either. In *Taking the Wheel*, Virginia Scharff explains that the motor vehicle was "born in a masculine manger, and when women sought to invade its power, they invaded a male domain."[7]

This is by way of reiterating that the most significant aspects of technology involve not internal dynamics but rather its relations with society and culture, how it figures in terms of *power* and *domain*. With highways and motor vehicles, it happens that the intersections involving transport, significant though they may be, are less freighted with meaning than in the realm of travel. Something Henry Adams wrote at the turn of the century is suggestive. Adams remarked that he "had been born with the railway system; had grown up with it; had been over pretty nearly every mile of it with curious eyes, and knew as much about it as his neighbors; but not there could he look for a new education. Incomplete though it was, the system seemed on the whole to satisfy the wants of society better than any other part of the social machine . . . . Nothing new had to be done or learned there, and the world hurried on to its telephones, bicycles, and electric trams."[8]

Mass Production

Railways were old hat, but Henry Adams could not help noticing bicycles and tramcars, the conveyances to which the world was hurrying for individual mobility and mass mobility. Actually, the future belonged to yet another kind of conveyance, the automobile. Like the steam locomotive, the auto was born overseas; the gasoline-fueled internal-combustion engine and the concept of applying it to a road vehicle both originated in Europe. But "the transformation of the automobile from a luxury for the few to a convenience for the many was definitely an American achievement, and from it flowed economic and social consequences of almost incalculable magnitude." So wrote John B. Rae on the first page of *The American Automobile* in 1965, and by that time the auto manufacturing industry was the key indicator of the state of the American economy.[9]

In recounting the history of travel and transport, it is with the automobile that one must first take extended note of *manufacture*. Conveyances of all sorts, large or small, had nearly always been produced with craft methods, built one at a time. But the auto was the product of a new system dating to the latter part of the 18th century, when manufacturing entered a phase based on the utilization of new sources of power, new kinds of machinery, and new ways of organizing work, including the mechanized production of articles from components that were identical and thus interchangeable. In the United States, new production methods were applied to a growing range of consumer products, ibcluding clocks, sewing machines, and bicycles. While the first autos were hand-built in Europe, within a few years they were being made with interchangeable parts in the United States of America. In 1908, three autos were sent to England, dismantled, and the parts mixed together. When reassembled, they started right up and ran perfectly. It happened that these autos were Cadillacs, but that same year, 1908, Henry Ford introduced the Model T, the vehicle that established the concept of "mass production." In other countries, people often spoke of *Fordism*.[10]

The autos Ford initially produced were priced as high as $1,500 and intended to be competitive with Buicks. But he had his eye on another market. While rival manufacturers had considered producing cars cheaply, Ford understood that one first had to design a car that would be suitable for a mass market in terms of simple operation and ease of repair, as well as low price—and then one had to rein in the cost of manufacture. And the way to do that, said Ford, was "to make one automobile like another automobile, to make them all alike, to make them come from the factory just alike—just like one pin is like another pin when it comes from the pin factory."[11]

Ford began to attain these ends after he decided to make Model Ts and nothing but Model Ts. At first, production costs remained too high to sell autos as cheaply as he wanted. Then, in 1913, his braintrust came up with the idea of the moving assembly line. Workers would remain at one station, performing simple tasks over and over; ultimately, a chassis would be assembled in 45 steps, and the man who put in a bolt often did not put on the nut. The numbers tell the story: In 1908, when Model Ts were priced at $850, Ford sold 5,986 of them; in 1916, when they were priced at $350, Ford sold 600,000. By 1920, half the autos in the world were Model Ts. All told, Ford produced more than 15 million "Tin Lizzies." Clay McShane notes that Sleepy John Estes would sing the praises of "the poor man's friend" because it enabled African Americans to get out of the south.[12]

In the last analysis, however, Ford was insufficiently attentive to changing cultural context. Except for electric starters and enclosed steel bodies, the design of the Model T did not change for two decades. Initially aimed at a nation of farmers, its popularity waned as rural America "got out of the mud" as the result of major highway legislation in 1916 and 1921. By the time Ford finally retooled in 1927, the demand for low-cost automobiles was being filled by second-hand sales of other makes, and after 1930 the American automotive market would be dominated for four decades not by Ford but by General Motors (GM), which sold cars on credit, something Ford refused to do, and also paid attention to "styling." Until 1970, Chevrolet outsold Ford every year but two, 1935 and 1945.[13]

General Motors became the most powerful corporation in the world, and the most imperious. GM sold the railways most of their diesels. GM was responsible for the demise of the form of transit that Henry Adams thought represented the future, electric trams, trolley cars—though not by the overtly conspiratorial means that many people imagine. The electric street railway was invented at the same time as the auto, as was the interurban electric railway, conceived as a convenient alternative to steam railways for short trips between city and town. It soon failed at the consumption junction. Downtown street railways survived longer, but ultimately could not stem the appeal of the auto, particularly after World War II and 15 years of depression and austerity.[14] (See Figure 24.)

Figure 24. This Los Angeles scene dates from 1948. Within two years, the electric railway line from downtown to Baldwin Park would be out of service, its right-of-way appropriated for the new Santa Ana Freeway, which—in theory if not actuality—would facilitate commuting by people like the man in his Oldsmobile convertible. T. H. Desnoyers photograph, courtesy of the Krambles Archive.

MASS MOTORIZATION

The most immediate ancestor of the automobile was devised in Germany by Karl Benz and Gottlieb Daimler, and by the 1890s auto manufacture was well established in Europe. Americans took about a decade to catch up. When one firm after another began stepping through the doorway, however, a gap opened and continued to widen. In 1913, the United States had 1 auto for every 77 people, in Great Britain the ratio was 1 for 165. By 1927, the ratio was 1:5 in the United States, 1:44 in Great Britain, and 1:196 in Germany. New limited-access highways, often called "parkways," were funded locally (Figure 25), and Congress was partial to highway interests long before the interstate system became the largest public-works project in history. As roads absorbed a growing proportion of tax monies, so did traffic police and emergency rooms. The urban panorama became studded with signals, meters, one-way streets, and parking lots, while new subdivisions often lacked sidewalks. In many cities, once-profitable mass transit systems were starting to languish by the 1930s. Eventually the owners bailed out. Yet, patronage and physical plant both continued to decline even under the administration of public agencies. The reality was that the automobile provided a level of convenience that no form of mass transit could match, and, more importantly, it conferred a heady illusion of mastery and control.[15]

Autos also facilitated suburbanization—pejoratively, sprawl—for it no longer seemed important to live close to traditional downtown amenities. In the suburbs, "drive-in" restaurants, theaters, and banks began to appear, even churches and liquor stores. Suburban flight was slowed by World War II, but since 1945 has never been reversed. Social critics lamented the waste of resources, the blight in central cities, the neglect of public transit. Downtown retailers had often been proponents of limited-access highways, which, they thought, would alleviate traffic congestion on city streets. But things did not work out that way; rather, the effect of new highways was centrifugal, for jobs, for housing, and for retail sales. For a growing number of Americans, whatever trade-offs automobility might entail were worth the price. And this was hardly just a male preference. Housewives and career women often seemed even more eager than men to embrace the auto, with its promise of privacy, personal security, and independence.[16]

Figure 25. One of the early limited-access highway projects in metropolitan Los Angeles was this stretch of Ramona Blvd., east of the city, completed in 1937. The electric railway at right was on borrowed time. Courtesy Matthew Roth, Automobile Club of Southern California.

By the 1960s, with the social costs of mass motorization becoming obvious, there were tentative efforts to implement technological fixes. Ralph Nader's *Unsafe at Any Speed* marked a new turn in the literature of motoring, a book that did *not* "celebrate automobiles, the companies that made them, and the men who made the companies," in Rudi Volti's words. The federal government began mandating standards for emissions control and safety. After the so-called oil crises of the 1970s, federal fuel-economy standards were enacted to discourage the production of "gas-guzzlers." A generation later, all seems to have been forgotten, as overpowered SUVs dominate suburban landscapes everywhere—just as they dominate the open road, for the great preponderance of intercity travel takes place in private motor vehicles. Railway trains account for less than 1 percent of the total. Except for a small market-share held by buses, all the rest takes place in airplanes, a mode of travel that long seemed exotic—before World War II, only a tiny proportion of the populace had ever been in an airplane. Air travel did not surpass travel by railway until the late 1950s. In 1960, the railways recorded two-thirds the passenger miles of the airways, but in 1970 only one-tenth.[17] (See Figure 26.)

Figure 26. In 1902, the New York Central's premier express train covered the 980 miles from New York to Chicago at an average speed of 49 miles an hour, and a locomotive such as this one, with very large driving wheels, was capable of 100. Sometimes an express locomotive was called a *flyer*; actual flight was science fiction.

9

Airways

A perennial debate among historians of technology concerns the question of whether there was anything inevitable about the combination of gasoline engines and the automobile. Was the situation potentially "malleable"? Might steam power, or storage batteries and electric motors, have worked just as well? Or better? What does it really mean to say that a device "works"? Works for whom? Is the hegemony of internal combustion due to intrinsic technological characteristics or simply the result of contested politics creating winners and losers? Those who argue for the latter often do so with an eye on the present, with the American gluttony for petroleum inflaming the clash of civilizations and with attempts to persuade consumers to step through the doors of new-age electric autos and hybrids not having much affect.[1]

The debate may persist, but rarely will one hear it extended to aeronautics. At the outset, battery- and steam-powered automobiles may have been on equal footing with internal combustion, perhaps a step ahead in some respects. But with an airplane, internal combustion was the only door open, the one device that could deliver sufficient power in a lightweight package. (The turbojet was a distant door, not yet ajar.) Which is by way of repeating a remark made at the outset—"real world constraints" on technology may trump everything else. Aeronautics involves a number of factors that lack social malleability, not the least of which is the phenomenon of gravity. At first, gravity seemed to render the notion of "heavier-than-air" flight an impossibility, like perpetual motion. Then, when flight did appear possible, gravity imposed rigorous bounds on technological design. With a wheeled vehicle, light weight might be desirable; with an airplane, it was essential.[2]

When it comes to the relationship of aeronautical technology with American culture, however, there is scarcely anything else in the history of transport and travel that has been as rich in choices that were contingent on social psychology and political context.

Technological Dreams

Dreams of flight figured prominently in Roger Bacon's *Secrets of Art and Nature* (1250), and were later foretold graphically by Leonardo da Vinci, whose "ornithopter" imparted a sense that if men were ever going to fly it would be with flapping wings that imitated nature. The power of this fallacy actually closed a door to human flight for several centuries. "Lighter than air" was a different matter, however. Balloonists were attracting attention in France by the late 18th century, and in the United States as many as 8,000 people had gone aloft even before the Civil War. Ballooning, so to say, was established as a fairly routine, fairly safe activity. But in their pursuit of private visions of powered, heavier-than-air flight, enthusiasts like Wilbur and Orville Wright knew they were risking their lives.[3]

The Wright brothers were bicycle mechanics in Dayton, Ohio, shrewd about practical matters but also well versed in technical literature and sophisticated about theory. They were not the first to realize that an airplane's wings had to be stationary in order to provide lift—that power had to come from another source—nor to perceive that existing internal-combustion engines could readily impel an "airframe" if properly designed. But nobody else had their appreciation of where the real technical challenges lay: flight would require a novel sort of propeller, not like marine propellers at all, and it would require means of control that nobody yet understood. Patiently and methodically, they solved the first problem with an eye on theoretical parameters, the second by experimenting with a glider that could be steered up or down, left or right, by manipulating horizontal and vertical control surfaces. On December 17, 1903, on the dunes of Kitty Hawk, North Carolina, they completed a series of four powered flights and not long afterward Wilbur flew several miles, making controlled turns.[4]

Yet, their feats attracted little attention. For one thing, hovering in the background was a controversy, but not a Fitch–Fulton sort of controversy involving differences of interpretation. Rather, it was a residue from the notorious failure of Samuel P. Langley's "aerodrome," which had been funded to the tune of $50,000 by the of War Department, plus more from the Smithsonian Institution. This was the institution Langley headed, but he was also "a government agent, supplied with government money for the purpose of perfecting a machine which the government might use as a substitute for its present armaments." Langley left no such legacy; instead, he left a rather widespread perception that human flight was a chimera, or, at best, a pretext by means of which well-connected scientific men could wangle public funds to pursue private enthusiasms. Many people simply would not accept the truth of accounts of what the Wright brothers were up to, in spite of their periodic test flights at a Dayton field that was in full view of anyone riding a trolley on a nearby line.[5]

Not until five years after Kitty Hawk, when the brothers staged a demonstration for the War Department at Fort Myer, Virginia, did they finally allay skepticism. Events at Fort Myer were widely publicized, and more publicity attended Teddy Roosevelt's "bully" venture aloft in 1910 (Figure 27) and the ensuing epoch in which aerial performers became standard fare at carnivals and county fairs. Many of these men and women—and indeed there were women, Blanche Scott, Mathilda Moisant, and Harriet Quinby, among others—met their death while stunting, and

Figure 27. In 1910, Theodore Roosevelt settles into a Wright biplane with pilot Arch Hoxsey looking on. The idea was to dispel notions of danger, but TR's flight was plenty worrisome; during the next year, more than a hundred fliers died in mishaps, and this was the fate that awaited Hoxsey himself. Courtesy National Air and Space Museum.

this tended to impart a sense that airplanes were awfully dangerous. But air shows had other significant effects as well: somewhat changing stereotypes of American womanhood, for one thing (Figure 28), and also helping spread the word about what Joseph Corn terms "the winged gospel."[6]

Contributing to this phenomenon were notable feats carried through without mishap, or at least without fatal mishap. In 1910, one of the Wrights' first competitors, Glenn Curtiss, won a $10,000 prize offered by a New York newspaper for flying from Albany to New York City, a feat reminiscent of Robert Fulton's a century earlier. The next year, Cal Rogers set out to fly from New York to Pasadena, in quest of a $50,000 prize offered by William Randolph Hearst. He crashed a dozen times and lost out on the prize because he took too long, but finally he made it, albeit bruised and battered. Every such event made headlines and enhanced the nation's "air-mindedness," an attitude Corn defines as "having enthusiasm for airplanes, believing in their potential to better human life, and supporting aviation development."[7]

Figure 28. Louise Thaden, whose day job was selling private planes, strikes a victorious pose after a race in 1929. Seven years later she won the Bendix Trophy, a coast-to-coast race in which women took three of the first five places. In her 1938 book, *High, Wide, and Frightened*, Thaden wrote that "flying is the only real freedom we are privileged to possess," and yet the cockpits of commercial airliners remained an exclusive male domain until the 1970s. Courtesy National Air and Space Museum.

Air-mindedness took on the trappings of a messianic religion. Claude Ryan, who later built the plane in which Charles Lindbergh crossed the Atlantic, remarked "that if he could talk anyone into his first flight, he would change that person's life for the better"—which is to say, he would "convert" that person. Indeed, people believed that the end of humankind's earthbound existence marked the turning point in the history of civilization. All the major religions posited a God who dwelled in a place to which one needed to fly, and a sense of flying as something divine was probably enhanced when "skywriting" first appeared over Seattle in 1913. As for more down-to-earth effects, the airplane was seen as a solution to the "railroad problem" and to many other problems as well. It foretold, said pioneer airplane manufacturer Jack Northrop, the possibility of "a glorious future of understanding and brotherhood."[8]

THE DARK SIDE

Even though the Wright brothers' Kitty Hawk *Flyer* remained a hallowed relic, utopian visions of a great leap forward began to fade in the latter 1930s. This had something to do with the reality that people still died in airplane crashes, including celebrities like Knute Rockne, Will Rogers, Wiley Post, and Amelia Earhart. Moreover, while private planes would become commonplace, utopian prospects of "an airplane for everyman" evaporated in the face of many practicalities. After World War II, dreams of world peace lay in ruins as a result of a new military strategy that used planes to incinerate whole cities. Later, the firm that Northrop founded would prosper by selling "warplanes" to almost anyone who could pay, and of course there was no sign whatsoever of any "glorious future of understanding and brotherhood."

The Wright brothers had shared a commonplace belief in the airplane's potential to render warfare obsolete, but as early as 1914 the Lafayette Escadrille began to cloak aerial combat in romantic garb. In 1917, Congress appropriated an unheard-of *$640 million* to supply the Allies with 20,000 planes. Even though airpower's role during the World War was largely in the realm of reconnaissance, some people hungered for a far bigger role in the future. During the interwar years, enthusiasts for military airpower constructed an enhanced mission for planes as a discrete kind of offensive weapon, and eventually this led to the development of the bombsight and the concept of strategic bombing. In our own time, this concept has been transformed into a sort of warfare that is supremely surreal: "invisible planes sent on their missions from scientifically advanced bases elsewhere to pick out unseen targets from high-tech screens and launch laser-guided or photo-guided weapons of destruction." These are the words of David Halberstam, who notes that during NATO's offensive against Serbian military units in Kosovo, pilots of B-2 stealth bombers would take off from their base in Missouri, fly halfway around the world to strike targets, and then the only question was "whether they would get back in time to watch their children's soccer and baseball games."[9]

POLITICS, TECHNOLOGY, AND COMMERCE

It bears noting that the cost of a B-2 bomber, two billion dollars, is about what the entire Tenn-Tom Waterway cost. Today, when one boards a commercial jetliner, one boards a conveyance whose design and engineering are mere spinoffs from vast appropriations for military R&D. And, just as there is no airplane for everyman, there has obviously been no escape from warfare, quite the contrary. Kitty Hawk had many unanticipated consequences. A century before, people had immediately seen that *The Clermont* and *The Best Friend of Charleston* were going to change habits of travel. But hardly anyone—not even the Wright brothers themselves—saw much likelihood of the airways becoming a major avenue of commercial travel. In 1910, when Glenn Martin inaugurated airplane manufacture in Southern California, few people imagined that this business would ever amount to much.

So, why did things turn out so differently? As usual, political considerations were transformative. Even before that first gigantic military appropriation in 1917, Congress had begun voting funds for airmail, and after the war the Post Office Department took direct responsibility for hiring pilots. Backed by its congressional partisans, the post office would drive demand for improved infrastructure, longer and smoother runways, floodlights, and navigational aids. By 1927, when a former airmail pilot, Charles Lindbergh, crossed the Atlantic nonstop and alone, airmail was a considered essential to major segments of the American economy such as banking. So was express package service.[10]

Direct government involvement in flying the mail seemed too socialistic for Calvin Coolidge's America, and did not last long. In 1925 Congress responded to pressure from would-be contractors flaunting the banner of free enterprise. Then, as contract carriers began to consolidate in order to serve longer routes, corporate names that are still familiar emerged: United, American, Delta, Northwest. Competition among such firms spurred the development of aircraft more suited to flying the mail, and these also began to incorporate seats for passengers, as many as ten of them in a Trimotor, Henry Ford's entry into this market. As new air-cooled powerplants from firms like Pratt and Whitney opened the door to commercial aircraft capable of "cruising" faster than 100 mph, more paying passengers showed up, 173,405 of them in 1929, 3.5 million in 1941.[11]

From the 1930s on, aviation history has traditionally been related in the language of triumphalism, and still is to a great degree—by legions of amateur historians, and even by professionals, many of whose narratives continue to echo the title of a standard book on airline industry, *Climb to Greatness*. For example, in a recent volume said by its editor to reflect a "new aviation history," William M. Leary writes that "all too often, political and parochial interests slowed the advance of the airline industry towards becoming a major factor in the nation's transportation system," as if some foreordained outcome were being thwarted by mere politics. Stephen McFarland writes that "the world wanted flight that went higher, faster and farther . . . because of the universal human compulsion for better performance and because the competitive urge demanded it."[12] In carrying forward the tenets of air-mindedness, these two historians might find a bone to pick with other scholars concerned with theoretical perspectives, or with the relationship between technology and engineering practice, or ideology, or culture. Rather than what "the world" might want, such historians seek to identify discrete human agents, and they are skeptical about the existence of "universal compulsions."[13]

But the fact remains that if one wants to think in terms of "a compulsion for better performance," the history of aviation yields plenty of material evidence. The Trimotor, which debuted in the 1920s, was followed by the radial engine in the early 1930s, four-engine planes in the late 1930s, and the turboprop and turbojet after World War II, when the federal government also began lavishing funds on the enhancement of infrastructure. In an essay fully attentive to the shortcomings of the "deeply deterministic" narratives that suffuse aviation history, Deborah Douglas describes the signal contributions made to the development of the modern airliner by the National Advisory Committee for Aeronautics (NACA), an agency established by Congress in 1915 "to supervise and direct the scientific study of the problems of flight." Alex Roland terms NACA "arguably the most important and productive aeronautical research establishment in the world."[14] Aluminum, a new material devised in the 1880s, finally found a major market. The public began to hear about aerodynamics, a technique influential far beyond aircraft engineering. Aerodynamics was also linked to aesthetics and invoked as an emblem of modernity. Under siege by the highway lobby and by automakers in the 1930s, railways brought aerodynamics into the design of passenger trains called "streamliners," marketing tools which aimed to convey an up-to-date image. It worked, for a while.

Perhaps not surprisingly, the story of American airways seems to recapitulate others that had already played out with different modes of transport and travel. The government provided help by various means while businesspeople were able to hide behind a free-enterprise facade. The pioneering epoch was marked by great danger—out of the 40 pilots hired for airmail service in 1920, only 9 were still alive when contract operation began five years later. Conveyances outdistanced infrastructure, at least until concerned parties could find their way into the public purse. Perhaps most importantly, speed yielded a crucial competitive edge, just as it once had done on seaways and railways.

Between the 1930s and the 1970s, the cruising speed of airliners doubled and tripled, and in that same interval air travel eclipsed travel by sea, and by rail too, though perhaps not with such apparent finality. The current share of intercity travel carried by railways is only 1 percent. This share could shrink even further, or it might grow, but only if political priorities shift in accord with the reality that the airline industry is to some extent "a public-works project."[15] The railways will have to be envisioned anew in terms of competitive advantages that outweigh speed. There is, for example, the towering annoyance of getting to a plane at Logan Airport in Boston, compared to the ease of stepping aboard a train at South Station. But the airplane remains a powerful political vehicle, and it remains to be seen whether contrasts like this can make a difference.

Conclusion

Speaking about the Los Angeles megalopolis, but in terms pertinent to the history of travel and transport in general, a distinguished professor of urban planning writes that major innovations in transportation policy "were presented as paragons of 'modernity,' packed with metaphors of innovation, technological change, and futurism. . . . The imagery of modernity has been consistently manipulated by politically powerful interests to suit their goals."[1] The success—the very existence—of many transportation technologies has depended on the way that people with their hands on the levers of political power have "manipulated" those levers, typically by providing overt or hidden subsidies, sometimes by finding ways to shoulder capital or developmental costs. Even when the enterprise has largely been private, as with clipper ships (and even clippers carried mail), there have always been those rhetorical flourishes of great psychological power: national honor, global strategy. The logic is often vulnerable, but we need to recall all the ways in which patriotic or martial or romantic urges can be "harnessed."

At the same time that Congress was voting to subsidize the Collins Line's oceanic speedsters in the mid-19th century, it also passed a special appropriation on behalf of Dr. Charles Grafton Page, to be used for building and testing a "galvanic" locomotive said to be capable of traversing the arid west without need of fuel or water: no firebox, no boiler. Instead, the power came from primary-cell batteries and an electric motor. In the words of Page's prime partisan, Senator Thomas Hart Benton of Missouri—a man who always had his eye on Manifest Destiny—such a locomotive could facilitate the establishment of a "great western road to India." Ultimately, trying to harness the forces of nature by consuming nitric and sulphuric acid in a battery instead of burning wood or coal in a firebox did not seem like sound science, and, after voting $20,000 for research and development, the Senate turned Page down when he asked for more.[2]

But, now, jump forward into the 20th century and consider a distant cousin of Page's galvanic locomotive, the electric tramcar, the trolley, the novel conveyance that caught Henry Adams's eye. In this case, there was no question about its practicality, and in Adams's time municipal authorities granted trolley magnates all sorts of special privileges. Some of these men became fabulously wealthy. (The Widener Library at Harvard, the Yerkes Observatory in Chicago, and the

Huntington Library and Art Gallery in California all stem from riches accumulated by men who owned trolley systems.) As more and more commuters opted for automobiles after World War I, however, trolleys fell on hard times, especially after the highway lobby became the new darling of officials able to confer special privilege. Condemned as hopelessly outmoded during the interwar years, trolleys were junked by the thousands after 1945 and within a generation they had virtually disappeared from the urban panorama, along with their infrastructure. A return seemed scarcely more likely than the return of canal boats to the countryside (Figure 29).

Figure 29. This pathetic scene made the pages of *Life* magazine in 1956: a Los Angeles scrapyard stacked with trolley cars that had been displaced by buses. Author's photo.

Yet, something of the sort *has* returned, and in considerable numbers, under the fashionable new guise of "light rail vehicles" (Figure 30). We have LRVs not because of any natural succession, but because urban transit, like all forms of transportation, is politically contingent. In 1936, for example, federal legislation forced electric utility companies to cast off their transit subsidiaries and left a weak industry adrift. Thirty years later—after an unremitting hard-sell by automakers—trolleys remained in operation in only a handful of North American cities. Was this a matter of evolution? Progress? Neither, insisted one vocal group of enthusiasts: it was a matter of conspiracy. The conspiratorial explanation for the demise of trolleys played again and again, in magazines like *Harpers*, on television's *60 Minutes*, during countless cocktail-party conversations. It was most colorfully recapitulated in the Steven Spielberg movie *Who Framed Roger Rabbit*, when the murderous Judge Doom revealed to detective Eddie Valiant his plan for transforming metropolitan Los Angeles. The year was 1947:

"They are calling it a freeway," says Doom.
"Freeway? What the hell's a freeway?" asks Valiant.
Doom grows agitated. "Eight lanes of shimmering cement running from here to Pasadena. Smooth. Safe. Fast. Traffic jams will be a thing of the past."

Figure 30. After trolley service between Los Angeles and Long Beach had been absent for thirty years, it was revived in 1991. Critics decried the billion-dollar outlay for elaborate infrastructure and modernistic conveyances—derived, in part, from a general sales tax—and the simultaneous neglect of mundane bus service essential to the needs of the poor and elderly. Author's photo

"Come on," says Valiant. "Nobody's going to drive this lousy freeway when they can take the Red Car for a nickel."
"Oh, they'll drive," Doom promises. "They'll have to. You see, I bought the Red Car so I could dismantle it."

For "Judge Doom," read General Motors. For "Red Car," read Pacific Electric, the PE, Southern California's storied network of electric railways. And indeed the Red Cars *were* gone soon enough, and GM *did* bear responsibility insofar as it relentlessly portrayed automobility as if it were the key to power, glory, and all of human fulfillment. But a conspiracy? A "noir fantasy," says Matthew Roth. True, the triumph of the motorcar was not exactly a matter of some "pure democratic impulse," but it was hardly possible that GM could conspire to "dismantle" the PE interurban system when it never controlled this system in any way. GM subsidiaries did purchase some streetcar companies, about 10 percent of the nation's total number, but substituted buses no more quickly, and sometimes with more deliberation, than those who owned the other 90 percent.[3]

Then, no sooner was the country down to just a handful of trolleys than the political context again began to shift in the face of concerns about congestion, resource allocation, and the environment. In the latter 1960s, Congress created an agency to foster Urban Mass Transit and made federal funds available for upgrading the few trolley lines that survived, and for new ones as well. A sort of conveyance that had long ago gained political favor and then lost it had, amazingly, *regained* favor! There soon developed a synergism among the strangest of bedfellows: liberal apostles of social engineering; right-wingers for whom the ascendence of automobility was not "a free market outcome"; technological enthusiasts for whom trolleys were a matter of deep personal affection; municipal boosters concerned about positive imagery; and sundry other parties with patently self-serving motives. The result: at the beginning of the 21st century, trolley cars, LRVs, are *back*—not only in Los Angeles, but also in several other California cities, and indeed in cities all across the land where they had been absent for more than a generation.[4]

Not that this is necessarily a bad thing. LRVs are of course environmentally "clean." But if the main problem is diesel buses with their lethal exhaust, the question is whether officials *must* step through the LRV door, instead of another open doorway, the door to nonpolluting buses, a far less expensive alternative. LRV lines in Los Angeles have cost about a billion dollars apiece, and certainly bespeak a city with a new face of modernity. But at the consumption junction some will lose as others gain. The gain comes to those who happen to want to go where the LRVs go. The losers are the vast majority of commuters, poor, nonwhite—if there is a "typical" bus patron, it is a female domestic—who are stuck with a bus system that is habitually neglected because of those showpiece electric railways: "Third-world buses for third-world people," says Mike Davis, known for his bleak portrait of the *City of Quartz*, and the head of the Transit Riders Union fights a desperate battle on behalf of the 90 percent for whom buses are the sole option.[5]

In the mid-1960s, there were streetcars operating in only six U.S. cities; LRVs can now be found in four times that many. But there is really no call for surprise about any such turn of events. At one point, President Bill Clinton floated a trial balloon about reviving the SST, the device that had so energized environmentalists when they succeeded in thwarting it on the basis of excessive cost and atmospheric degradation. Nobody should have felt certain that trolley cars would be gone forever, and nobody ought to dismiss the possible revival of the SST, somehow, no matter how irrational it may seem. And lest I leave the impression that this sort of irrationality is exclusively American: not to forget that a British–French consortium actually *did* go ahead with a supersonic transport, the *Concorde*, which has flown for three decades, gushing red ink, an airborne monument to the power of political appeals to cosmic destiny.

Or, consider China's Ji-Tong Railway, which operates the last main-line steam locomotives in the world but soon intends to replace them with diesels, at great cost. Why, because they are worn out? No, many are nearly new. Because diesel oil is cheaper? No, there are vast coalfields in this region of Inner Mongolia. Because steam is not viable for some other reason? No, officials concede that it is viable. But, as with railways in the Soviet Union (which, embarrassingly for the Kremlin, still had steam locomotives when steam had all but disappeared from North America) the phase-out is rationalized on the grounds that steam does not accord with "the modern socialist image."[6]

So often, technological change hinges on the perceptions of politically powerful people who take matters like "image" seriously. In the 1850s such people rendered it feasible for Edward K. Collins to build an ocean liner capable of 18 knots, fastest in the world. A century later, such people deemed it worth nearly any price to build a liner that could cross the Atlantic in less than three and a half days, again the fastest in the world. Now, because one can look through a figurative doorway and envision futuristic magnetic-levitation trains, we may actually see maglev on high-profile routes even while vast reaches of our conventional railway network have no passenger service at all. Or we may see well-publicized stretches of "automated" highway—vehicles steered by computers—even as other portions of the highway system slip into decrepitude. If one can imbue the keepers of the public purse with sufficient enthusiasm for novelty or sufficient faith in progress, scarcely any dream seems too grand. Technological change is often inexplicable in terms of straightforward necessity. If it always were, studying the history of technology would not be nearly as instructive as it is.

NOTES

INTRODUCTION

1. On technology and agency, see Leo Marx and Merritt Roe Smith, eds., *Does Technology Drive History: The Dilemma of Technological Determinism* (Cambridge, Mass.: MIT Press, 1995). "New device merely opens a door" is quoted from Lynn White Jr., *Medieval Technology and Social Change* (Oxford, U.K.: Oxford University Press, 1962), 28. Counter to White's remark, an argument that a particular technology, computers, now "drives human beings as an unstoppable force" is set forth by Rosalind Williams in "'All That is Solid Melts into Air': Historians of Technology and the Information Revolution," *Technology and Culture* 41 (2000): 641–668, quote on 642. Amending White, Melvin Kranzberg notes that "an open door is an invitation" ("Technology and History: 'Kranzberg's Laws,'" *Technology and Culture* 27 [July 1987]: 545). Two expressions used here were given currency in the title of Langdon Winner's *Autonomous Technology: Technics-out-of-Control as a Theme in Political Thought* (Cambridge, Mass.: MIT Press, 1977).

2. "Somewhere to nowhere" quote is from Carroll Pursell, *The Machine in America: A Social History of Technology* (Baltimore, Md.: Johns Hopkins University Press, 1995), 83. On motivations for invention, a useful analysis is Joseph Rossman, *Industrial Creativity: The Psychology of the Inventor* (New York: University Books, 1964); on rationality, see "'The Frailties and Beauties of Technological Creativity': An Interview with John M. Staudenmaier by Robert C. Post," *Invention and Technology* (spring 1993): 16–24. On enthusiasm, see Eugene Ferguson, "Toward a Discipline of the History of Technology," *Technology and Culture* 14 (1974): 13–30, and Robert C. Post, "Technological Enthusiasm," *The Facts on File Encyclopedia of Science, Technology, and Society* ed. Rudi Volti (New York: Facts on File, 1999), 999–1001. For a case study, see Post, "Strip, Salt, and Other Straightaway Dreams," in *Possible Dreams: Enthusiasm for Technology in America*, ed. John Wright (Dearborn, Mich.: Henry Ford Museum and Greenfield Village, 1992), 98–109. On the failure to anticipate responses to the auto, suggestive is Edward Tenner, *Why Things Bite Back: Technology and the Revenge of Unintended Consequences* (New York, Vintage Books, 1996), xi; or, on the failure to anticipate it at all, Dave Walter, ed., *Today Then: 1993 as Predicted in 1893* (New York: American and World Geographic Publishing, 1993).

3. Reagan quoted in *Los Angeles Times*, 20 May, 1971. One gets a sense of the political influences on technological change from another Reagan remark about the SST being essential "to the future of thousands of California aircraft industry workers." On Moynihan and maglev, see Bruce E. Seely, "Back to the Future? High-Speed Rail and Historical Patterns of American Transportation

Development," *Railroad History* 170 (spring 1994): 5–14. On light rail, Robert C. Post, "Urban Railways Redivivus: Image and Ideology in Los Angeles, California," in *Suburbanizing the Masses: Public Transport and Urban Development in Historical Perspective*, ed. Colin Divall (York, U.K.: National Railway Museum, 2002).

4. Allan Sloan, "Planes, Trains and Politicians," *Newsweek*, 2 October, 2002, 49.

5. Leo Marx, *The Machine in the Garden: Technology and the Pastoral Ideal in America* (New York: Oxford University Press, 1964). Ruth Schwartz Cowan, "The Consumption Junction: A Proposal for Research Strategies in the Sociology of Technology," in *The Social Construction of Technological Systems*, eds. Wiebe E. Bijker et al. (Cambridge, Mass.: MIT Press, 1987), 261–280; Cowan, "From Virginia Dare to Virginia Slims: Women and Technology in American Life," *Technology and Culture* 20 (1979): 51–63, quote about "the relation" on 51. For studies in how this may differ, see Virginia Scharff, *Taking the Wheel: Women and the Coming of the Motor Age* (New York: The Free Press, 1991); Margaret Creighton, "Sailing between a Rock and a Hard Place: Navigating Manhood in the 1800s," in Benjamin W. Labaree et al., *America and the Sea: A Maritime History* (Mystic, Conn.: Mystic Seaport, 1998), 293–296; Ruth Oldenziel, *Making Technology Masculine: Men, Women, and Modern Machines in America, 1870–1945* (Amsterdam: Amsterdam University Press, 2000); Julie Wosk, *Women and the Machine: Representations from the Spinning Wheel to the Electronic Age* (Baltimore, Md.: Johns Hopkins University Press, 2001); and Roger Horowitz, ed., *Boys and their Toys: Masculinity, Class, and Technology in America* (New York: Routledge, 2001). On agency, Ronald Kline and Trevor Pinch, "Users as Agents of Technological Change: The Social Construction of the Automobile in the Rural United States," *Technology and Culture* 37 (1996): 763–795.

6. On complex hierarchies, see Alfred D. Chandler Jr., *The Visible Hand: The Managerial Revolution in American Business* (Cambridge, Mass.: Harvard University Press 1977); I say "assumed to have engendered" because an argument to the contrary is set forth by Gerald Berk in *Alternative Tracks: The Constitution of the American Industrial Order, 1865–1917* (Baltimore: Johns Hopkins University Press, 1994). The remark about the Camden & Amboy is from Henry Varnum Poor, *Manual of Railroads and Canals* (1860), quoted in Harry Sinclair Drago, *Canal Days in America* (New York: Brimhall House, 1972), 296. Data on land grants is from John F. Stover, *The Routledge Historical Atlas of American Railroads* (New York: Routledge, 1999), 32–33. As indicated in the suggestions for further reading, there are numerous surveys of American railroad history, but for the nonspecialist this atlas is accurate and concise.

7. Lynn White Jr., "The Historical Roots of Our Ecological Crisis," reprinted from *Science*, in *Philosophy and Technology: Readings in the Philosophical Problems of Technology*, ed. Carl Mitcham and Robert Mackey (New York: Free

Press, 1983), 259–265. Daniel Headrick, *Tools of Empire: Technology and European Imperialism in the Nineteenth Century* (New York: Oxford University Press, 1981). Mel Horwitch, *Clipped Wings: The American SST Conflict* (Cambridge, Mass.: MIT Press, 1982). "A Dangerous Addiction," *The Economist*, 15 December, 2001.

8. Steven W. Usselman quoted from *Regulating Railroad Innovation: Business, Technology, and Politics in America, 1840–1920* (Cambridge, U.K.: Cambridge University Press, 2002), 5. The quote on the waterway is from the *Atlanta Journal and Constitution*, in Jeffrey K. Stine, *Mixing the Waters: Environment, Politics, and the Building of the Tennessee–Tombigbee Waterway* (Akron, Ohio: University of Akron Press, 1993), 249.

9. "Use of particularity" quote is from John Lauritz Larson, *Bonds of Enterprise: John Murray Forbes and Western Development in America's Railway Age*, rev. ed. (Iowa City: University of Iowa Press, 2001), xiii.

1. THE WESTERN OCEAN

1. On Asian migration, a useful synopsis is presented in Richard B. Morris, ed., *Encyclopedia of American History*, rev. ed. (New York: Harper & Row, 1965), 3–13.

2. The most recent scholarship on Ptolemy is presented in J. Lennart Berggren and Alexander Jones, *Ptolemy's Geography: An Annotated Translation of the Theoretical Chapters* (Princeton, N.J.: Princeton University Press, 2000). On the Norsemen, see Samuel Eliot Morison, *The European Discovery of America: The Northern Voyages, A.D. 500–1600* (New York: Oxford University Press, 1971), 32–80. Morison's work on exploration is like Francis Parkman's: dated and yet indispensable.

3. John Law, "Technology and Heterogeneous Engineering: The Case of Portuguese Expansion," in *The Social Construction of Technological Systems* ed. Wiebe E. Bijker et al. (Cambridge, Mass.: MIT Press, 1989), 111–134; also, Thomas P. Hughes, "The Evolution of Large Technological Systems," in ibid., 51–82.

4. Lynn White Jr., "Technology in the Middle Ages," in *Technology in Western Civilization*, ed. Melvin Kranzberg and Carroll W. Pursell Jr. (New York: Oxford University Press, 1967), 1:66–79, esp. 74–77; A. Rupert Hall, "Early Modern Technology, to 1600," in ibid., 79–103, esp. 95. Law quoted from "Technology and Heterogeneous Engineering," 118. S. C. Gilfillan's, *Inventing the Ship* (Chicago: Follett Publishing, 1935) is a provocative analysis by a sociologist of another stripe than Law.

5. Samuel Eliot Morison, *The Great Explorers: The European Discovery of America* (New York: Oxford University Press, 1978), 26–28, 393–394; Dava Sobel, *Longitude* (London: Fourth Estate, 1996).

6. Morison, *The Great Explorers*, 351–547.

7. Charles E. Nowell, *The Great Discoveries and the First Colonial Empires*, rev. ed. (Ithaca, N.Y.: Cornell University Press, 1965), 55–56.

8. On the transatlantic migration of the Puritans, see David Cressy, "'The Vast and Furious Ocean': Shipboard Socialization and the Atlantic Passage," chap. 6 in *Coming Over: Migration and Communication between England and New England in the Seventeenth Century* (Cambridge, U.K.: Cambridge University Press, 1987).

9. Captaine John Smith, *The General Historie of Virginia, New England & The Summer Isles*, 2 vols. (Glasgow, Scotland: James MacLehose and Sons, 1907), 2:12.

10. Fr. Font quoted in John Walton Caughey, *California* (New York: Prentice Hall, 1940), 156. "Redeem a wilderness" is quoted from Frederick Merk, *Manifest Destiny and Mission in American History: A Reinterpretation* (New York: Knopf, 1963), 130.

2. A New Nation

1. G. Terry Sharrer, "Naval Stores, 1781–1881," in *Material Culture of the Wooden Age*, ed. Brooke Hindle (Tarrytown, N.Y.: Sleepy Hollow Press, 1981), 241–270; Brooke Hindle, *Technology in Early America: Needs and Opportunities for Study* (Chapel Hill: University of North Carolina Press, 1966), 53. Americans also comprised much of the crew for the British merchant fleet, a situation fraught with political complications—the "impressment" of seamen being a major cause of the War of 1812.

2. Douglass C. North, *The Economic Growth of the United States, 1790–1860* (New York: Norton, 1966), 17–23.

3. On early road travel in brief, see Carroll Pursell, *The Machine in America: A Social History of Technology*, (Baltimore, Md.: Johns Hopkins University Press, 1995), 67–72, and William L. Richter, *Transportation in America* (Santa Barbara, Calif.: ABC–Clio, 1995), 429–437. On the harness, horseshoe, and whippletree, see Lynn White Jr., "Technology in the Middle Ages," in *Technology in Western Civilization*, ed. Melvin Kranzberg and Carroll W. Pursell Jr. (New York: Oxford University Press, 1967), 1:74–75; on the wheel in brief, Rudi Volti, ed., *The Facts on File Encyclopedia of Science, Technology, and Society* (New York: Facts on File, 1999), 1127–1128. One must not assume that the wheel represented technological *inevitability*: in *The Camel and the Wheel* (Cambridge, Mass.: Harvard University Press, 1975), Richard W. Bulliet explains how Middle Eastern and North African cultures first adopted wheeled vehicles and then abandoned them after devising saddles that enabled camels to be used as pack animals.

4. Pursell, *The Machine in America*, 69.

5. Ruth Schwartz Cowan quoted from *A Social History of American Technology* (New York: Oxford University Press, 1997), 94.

6. Philip D. Jordan, *The National Road* (Indianapolis, Ind.: Bobbs Merrill, 1948). On suspicion of centralized authority, see Bernard Bailyn, *The Ideological Origins of the American Revolution*, 2d ed. (Cambridge, Mass.: Harvard University Press, 1992), 56.

7. On road-building, see Dirk Struik, *Yankee Science in the Making* (Boston: Little, Brown, 1948), 98–112, and Don Berkebile, "Wooden Roads," in *Material Culture of the Wooden Age*," ed. Hindle, 129–158. Still worthwhile, too, is Seymour Dunbar, *A History of Travel in America*, 4 vols. (Indianapolis, Ind.: Bobbs-Merrill, 1915).

8. On Macadamized roads, see Richter, *Transportation in America*, 295, 430–432.

9. Barrie Trinder, *The Iron Bridge* (Telford: Ironbridge Gorge Museum Trust, 1979). The "public spirit" quote is in Struik, *Yankee Science in the Making*, 104. Frances C. Robb, "First and Still There," *Invention and Technology* (Fall 1994) 17, tells the story of the first iron bridge in the United States, an 80-foot span at Dunlap Creek in southwestern Pennsylvania on the National Road.

10. Trinder quoted from *The Blackwell Encyclopedia of Industrial Archaeology* (Cambridge, Mass.: Basil Blackwell, 1992), 778. See also S. B. Hamilton, "Bridges," in Charles Singer et al., *A History of Technology*, (New York: Oxford University Press, 1957), 3:417–436, and Eric DeLony, "The Golden Age of the Iron Bridge," *Invention and Technology* (Fall 1994) 8–22.

11. Struik, *Yankee Science in the Making*, 102–112. Richard G. Wood, *Stephen Harriman Long, 1784–1864: Army Engineer, Explorer, Inventor* (Glendale, Calif.: Arthur H. Clark Co., 1966). Wooden boxcars also had trussed sides, with rods and turnbuckles underneath to take up sag; "riding the rods" was a fabled mode of travel for hobos.

12. Mark Aldrich, "Engineering Success and Disaster: American Railroad Bridges, 1840–1900," *Railroad History* 180 (spring 1999): 31–72. John A. Kouwenhoven, "The Designing of the Eads Bridge," *Technology and Culture* 23 (1982): 535–68. On the Lucin Cutoff, see Freeman Hubbard, *Encyclopedia of North American Railroading* (New York: McGraw–Hill, 1981), 208, a satisfactory sketch of this project in a book that is not, however, always trustworthy.

13. Harry N. Scheiber, "The Transportation Revolution and American Law: Constitutionalism and Public Policy," in *Transportation in the Early Nation* (Indianapolis: Indiana Historical Society, 1982), 1–29; Douglas E. Clanin, "Internal Improvements in National Politics," in ibid., 30–60; George Rogers Taylor, *The Transportation Revolution, 1815–1860* (New York: Harper and Row, 1968), 15–31; Raymond H. Pulley, "Andrew Jackson and Federal Support of Internal Improvements: A Reappraisal," *Essays in History* 9 (1963–1964):

48–59. W. Turrentine Jackson, *Wagon Roads West: A Study of Federal Road Surveys and Construction in the Trans-Mississippi West, 1848–1869* (New Haven, Conn.: Yale University Press, 1965); Oscar O. Winther, "The Persistence of Horse-Drawn Transportation in the Trans-Mississippi West, 1865–1900," in *Probing the American West*, K. Ross Toole et al. (Santa Fe: Museum of New Mexico Press, 1962), 42–48.

14. On the Panama Railroad in brief, see Bob Johnston, "The First Transcontinental in America Didn't Go Through Promontory," *Trains*, September 2002, 46–47.

15. For a discussion of the "strong programme" in the sociology of knowledge, which holds that no part of a technological equation is "exempt from sociological explanation," see Donald MacKenzie, "How Do We Know the Properties of Artefacts? Applying the Sociology of Knowledge to Technology," in *Technological Change: Methods and Themes in the History of Technology*, ed. Robert Fox (Amsterdam: Harwood, 1996), 247– 263, quote on 260. For an analysis of "non-negotiable" elements in technological equations, Walter G. Vincenti, "The Technical Shaping of Technology: Real World Constraints and Technical Logic in Edison's Electric Light System," *Social Studies of Science* 25 (1995): 553–574.

3. SEAWAYS

1. On the China trade in brief, see Samuel Eliot Morison, *The Oxford History of the American People* (New York: Oxford University Press, 1965), 284, and Benjamin W. Labaree et al., *America and the Sea: A Maritime History* (Mystic, Conn.: Mystic Seaport, 1998), 158–159. On the value of imports and exports, Richard B. Morris, ed., *Encyclopedia of American History* rev. ed. (New York: Harper & Row, 1965), 515.

2. Howard I. Chapelle, *The Baltimore Clipper: Its Origins and Development* (Salem, Mass.: Marine Research Society, 1930).

3. On packets in brief, see Howard I. Chapelle, *The National Watercraft Collection* 2d ed. (Camden, Maine: International Marine Publishing Co., 1976), 26–29; for detail, Robert Albion, *Square-Riggers on Schedule: The New York Sailing Packets to England, France, and the Cotton Ports* (Princeton, N.J.: Princeton University Press, 1938).

4. Albion, *Square-Riggers on Schedule*; David Budlong Tyler, *Steam Conquers the Atlantic* (New York: D. Appleton-Century Co., 1939), 78–91.

5. Howard I. Chapelle, "The Pioneer Steamship *Savannah*," *United States National Museum Bulletin* 228 (1961): 61–80. On exports, Morris, ed., *Encyclopedia of American History*, 515.

6. Tyler, *Steam Conquers the Atlantic*, 123.

7. Tyler, *Steam Conquers the Atlantic*, 181–182. Robert C. Post, "The Thrall of the Blue Riband," *Invention and Technology* (winter 1996) 8–19. A knot is equal to one nautical mile (6,080 feet) per hour, or 1.15 miles per hour.

8. Tyler, *Steam Conquers the Atlantic*, 217–236. Eugene S. Ferguson, "John Ericsson and the Age of Caloric," *United States National Museum Bulletin* 228 (1961): 41–60.

9. Tyler, *Steam Conquers the Atlantic*, 239–241. On the inception of the *America*'s Cup, see Charles H. Jenrich, "'Flatboards' Won the First One," in *Yankees Under Sail: A Collection of Best Sea Stories from* Yankee *Magazine* ed. Richard D. Heckman (Dublin, N.H.: Yankee, Inc., 1968), 194–201.

10. On clipper ships in brief, see Alexander Laing, *American Ships* (New York: American Heritage Press, 1971), 185–213; Labaree et al., *America and the Sea*, 409; Peter Kemp, ed., *The Oxford Companion to Ships and the Sea* (New York: Oxford University Press, 1976), 172–173; and Nicholas Dean, "The Brief, Swift Reign of the Clippers," *Invention and Technology* (fall 1989) 48–55. For detail, Octavius T. Howe and Frederick C. Matthews, *American Clipper Ships, 1833–1858*, 2 vols. (Salem, Mass: Marine Research Society, 1926, 1927).

11. Carl C. Cutler, *Greyhounds of the Sea: The Story of the American Clipper Ship* (New York: G. P. Putnam's Sons, 1930).

12. On Down Easters in brief, see Laing, *American Ships*, 393–398; for detail, A. Basil Lubbock, *The Down Easters: American Deep-Water Sailing Ships, 1869–1929*, 2d ed. (Glasgow, Scotland: J. Brown & Son, 1930). On schooners, Laing, *American Ships*, 398–433; J. Lewis Parker, *The Great Coal Schooners of New England, 1870–1909* (Mystic, Conn.: Mystic Seaport, 1948). On the *Lawson*, Laing, 433–444, and Edward L. Rowe, as told to Capt. William P. Coughlin, "The *Lawson*'s First and Last Voyage," in *Yankees Under Sail*, ed. Heckman, 140–145.

13. Morris, ed., *Encyclopedia of American History*, 516. Goldenberg quoted from "With Saw and Axe and Auger: Three Centuries of American Shipbuilding," in *Material Culture of the Wooden Age*, ed. Brooke Hindle (Tarrytown, N.Y.: Sleepy Hollow Press, 1981), 97–128, on 97. McPhee quoted from *Looking for a Ship* (New York: Farrar Strauss Giroux, 1990), 11.

14. Edward R. Crews, "The Big Ship," *Invention and Technology* (spring/summer 1990) 34–41; Post, "The Thrall of the Blue Riband."

4. CANALS

1. Struik quoted from *Yankee Science in the Making* (Boston: Little, Brown and Co., 1948), 112. Mark Derr, "Network of Waterways Traced to Ancient Florida Culture," *New York Times*, 23 July, 2002. Francesca Bray, *Technology and Society in Ming China (1368–1644)* (Washington, D.C.: American Historical Association/Society for the History of Technology, 2000), 20–21. For the Canal

du Midi, see Chandra Mukerji, "Cartography, Entrepreneurialism, and Power in the Reign of Louis XIV: The Case of the Canal du Midi," in *Merchants and Marvels: Commerce, Science, and Art in Early Modern Europe*, ed. Pamela H. Smith and Paula Findlen (New York: Routledge, 2002), 248–276.

2. Alexander Crosby Brown, *Juniper Waterway: A History of the Albemarle and Chesapeake Canal* (Charlottesville: University Press of Virginia, 1981).

3. Ruth Schwartz Cowan, *A Social History of American Technology* (New York: Oxford University Press, 1997), 100–101. Cecelsik quoted from *The Waterman's Song: Slavery and Freedom in Maritime North Carolina* (Chapel Hill: University of North Carolina Press, 2001), 109.

4. Elting Morison, *From Know-How to Nowhere: The Development of American Technology* (New York: Basic Books, 1974), 32–34. Struik, *Yankee Science in the Making*, 114–119. Mary Stetson Clark, *The Old Middlesex Canal* (Melrose, Mass.: The Hilltop Press, 1974). Harry Sinclair Drago, *Canal Days in America* (New York: Brimhall House, 1972), 9–12.

5. Christopher Baer, *Canals and Railroads of the Mid-Atlantic States, 1800–1860* (Wilmington, Del.: Regional Economic Research Center, Eleutherian Mills–Hagley Foundation, 1981), 2, is by far the best regional survey. Ronald E. Shaw, *Canals for a Nation: The Canal Era in the United States, 1790–1860* (Lexington: University Press of Kentucky, 1990), is particularly good on the role of self taught engineers.

6. John Tarkov, "Engineering the Erie Canal," *Invention and Technology* (summer 1986) 50–57. See also Ronald E. Shaw, *Erie Water West: A History of the Erie Canal, 1792–1854* (Lexington: University of Kentucky Press, 1966), and a book from 30 years later, Carol Sheriff, *The Artificial River: The Erie Canal and the Paradox of Progress, 1817–1862* (New York: Hill & Wang, 1996). The former is triumphalist; the latter, in the words of John Lauritz Larson, "exposes the ambivalence and transformational dilemmas experienced by all parties when confronted with a major transportation 'improvement'" (*Bonds of Enterprise: John Murray Forbes and Western Development in America's Railway Age*, rev. ed. [Iowa City: University of Iowa Press, 2001], xiii).

7. Morison, *From Know-How to Nowhere*, 35–39.

8. "Based on the success" quote is from Frederick C. Gamst, ed., David J. Diephouse and John C. Decker, trans., *Early American Railroads: Franz Anton von Gerstner's* Die innern Communicationen *(1842–1843)* (Stanford, Calif.: Stanford University Press, 1997), 47, easily the most valuable published primary source on early canals as well as railroads. The classic synthesis is George Rogers Taylor, *The Transportation Revolution, 1815–1860* (New York: Harper and Row, 1968). On the Wabash and Erie, see Ralph D. Gray, "The Canal Era in Indiana," in *Transportation in the Early Nation* (Indianapolis: Indiana Historical Society, 1982), 113–134, and Ronald E. Shaw, "The Canal Era in the Old Northwest," in ibid., 89–112.

9. Cowan, *A Social History of American Technology*, 105.

10. On the C & O in brief, see Drago, *Canal Days in America*, 47–73; for detail, Walter S. Sanderlin, *The Great National Project: A History of the Chesapeake and Ohio Canal* (Baltimore, Md.: Johns Hopkins University Press, 1946), and Harold Skramstad, "The Georgetown Canal Incline," *Technology and Culture* 10 (1969): 549–560. On the Main Line, see Darwin H. Stapleton, "Moncure Robinson: Railroad Engineer, 1828–1840," in *Benjamin Henry Latrobe & Moncure Robinson: The Engineer as Agent of Technological Transfer*, ed. Barbara E. Benson (Greenville, Del.: Eleutherian Mills Historical Library, 1972), 33–60, and Robert McCullough and Walter Leuba, *The Pennsylvania Main Line Canal* (York, Pa.: American Canal and Transportation Center, 1973). On the cost of the entire Pennsylvania system, William L. Richter, *Transportation in America* (Santa Barbara, Calif.: ABC–Clio, Inc., 1995), 389.

11. Douglas E. Clanin, "Internal Improvements in National Politics, 1816–1830," in *Transportation and the Early Nation*, 30–60.

12. On the anthracite canals, see John H. Hoffman, "Anthracite in the Lehigh Region of Pennsylvania, 1820–1845," in *United States National Museum Bulletin* 252 (1968): 92–141, and Robert M. Vogel, *Roebling's Delaware and Hudson Canal Aqueducts* (Washington, D.C.: Smithsonian Institution Press, 1971).

5. An Inland Seacoast?

1. The interaction of technology and culture on Michigan's Upper Peninsula is analyzed by Larry D. Lankton in *Cradle to Grave: Life, Work, and Death at the Lake Superior Copper Mines* and *Beyond the Boundaries: Life and Landscape at the Lake Superior Copper Mines, 1840–1875* (New York: Oxford University Press, 1991, 1997). On specialized ore-carriers, notably the so-called whaleback, see Benjamin W. Labaree et al., *America and the Sea: A Maritime History* (Mystic, Conn.: Mystic Seaport, 1998), 374–375. The oldest known power plant from a propeller-driven steamship—the *Indiana*, launched in 1848, lost in 120 feet of water off Whitefish Point in Lake Superior in 1858—was retrieved in 1979 by the Smithsonian Institution in cooperation with the Army Corps of Engineers; see Robert C. Post, "*Indiana* Salvage: Oldest American Marine Engine Recovered," *Society for Industrial Archeology Newsletter* 7 (September 1979): 1+.

2. Jacques Lesstrang, *Seaway: The Untold Story of North America's Fourth Seacoast* (Seattle, Wash.: Salisbury Press, 1976).

3. Daniel J. McConville, "Seaway to Nowhere," *Invention and Technology* (fall 1995): 34-44.

4. "To hell with economics" is quoted in William L. Richter, *Transportation in America* (Santa Barbara, Calif.: ABC–Clio, 1995), 463; "an anachronism" is quoted from McConville, "Seaway to Nowhere," 43.

6. Riverways

1. Rudi Volti, ed., *The Facts on File Encyclopedia of Science, Technology, and Society* (New York: Facts on File, 1999), 148–149; David McCullough, *The Great Bridge* (New York: Avon Books, 1972).

2. On difficulties of navigating the Hudson, see Robert G. Albion, *The Rise of New York Port (1815–1860)* (New York: Charles Scribner's Sons, 1939). Law is quoted from "Theory and Narrative in the History of Technology: Response," *Technology and Culture* 32 (1991): 383. On the role of chronicles in priority disputes, see Charles C. Gillespie, *The Edge of Objectivity* (Princeton, N.J.: Princeton University Press, 1960), 218.

3. On the inception and development of steam technology in brief, see Paul F. Johnston, *Steam and the Sea* (Salem, Mass.: Peabody Museum of Salem, 1983), 12–34; on European antecedents, Robert C. Post, "Steamboats, " in Volti, ed., *The Facts on File Encyclopedia of Science, Technology, and Society*, 948–950.

4. C. M. Harris, "The Improbable Success of John Fitch," *Invention and Technology* (winter 1989): 24–31; on Rumsey, see Edwin T. Layton Jr., "'The Most Original'," *Invention and Technology* (spring 1987): 50–56.

5. Frank D. Prager, ed., *The Autobiography of John Fitch* (Philadelphia: American Philosophical Society, 1976). James Thomas Flexner, *Steamboats Come True: American Inventors in Action* (Boston: Little, Brown & Co., 1944), 267, 333–335. Rarely has a biography been more aptly titled than Thomas Boyd's *Poor John Fitch* (New York: G. P. Putnam's Sons, 1935).

6. Flexner, *Steamboats Come True*, 280–295; Louis C. Hunter, *Steamboats on the Western Rivers: An Economic and Technological History* (Cambridge, Mass.: Harvard University Press, 1949). For Hunter's line of analysis in brief, see "The Invention of the Western Steamboat," *Journal of Economic History* 3 (1943): 202–220.

7. Brooke Hindle, *Emulation and Innovation* (New York: New York University Press, 1981); John H. White Jr., "Robert Fulton's Dream," *Invention and Technology* (summer 2002): 39–46. One should also compare the tenor of a Fulton hagiography (a visit to amazon.com suggests that there are dozens of them) with that of the neo-Luddite Kirkpatrick Sale, *The Fire of Genius: Robert Fulton and the American Dream* (New York: Free Press, 2001), set against "the darker side of the American dream."

8. The Supreme Court case, *Gibbons v. Ogden*, is outlined in Richard B. Morris, ed., *Encyclopedia of American History* (New York: Harper & Row, 1965), 489. On the compelling notion of "a technology of haste," see Daniel Boorstin, "Getting There First," chap. 14 in *The Americans: The National Experience* (New York: Vintage Books, 1965).

9. Hunter, *Steamboats*, 217.

10. Leland D. Baldwin, *The Keelboat Age on Western Waters* (Pittsburgh, Pa.: University of Pittsburgh Press, 1941). Hunter, *Steamboats*, 22, 27 (quoting from *Annales Maritimes et Coloniales*, 1840).

11. Hunter, *Steamboats*, 6–20, 61–120, quote on 64. On Evans, see Eugene Ferguson, *Oliver Evans, Inventive Genius of the American Industrial Revolution* (Wilmington, Del.: Hagley Museum, 1980).

12. On getting results we do not want along with those we do, see David Pye, *The Nature and Aesthetics of Design* (London: Barrie & Jenkins, 1978), 18; also, Edward Tenner, *Why Things Bite Back: Technology and the Revenge of Unintended Consequences* (New York, Vintage Books, 1996). On explosions, see John G. Burke's classic "Bursting Boilers and the Federal Power," *Technology and Culture* 7 (1966): 1–23.

13. Hunter, *Steamboats*, 254–255, 272–277. On so-called mountain steamboats, which navigated to the very edge of the Rockies on the east and into Idaho on the west, see Carlos Arnaldo Schwantes, *Long Day's Journey: The Steamboat and Stagecoach Era in the Northern West* (Seattle: University of Washington Press, 1999), 124–127.

14. Hunter, *Steamboats*, 523–540. Inspection later moved to the Commerce Department.

15. Hunter, *Steamboats*, 543; Benjamin W. Labaree et al., *America and the Sea: A Maritime History* (Mystic, Conn.: Mystic Seaport, 1998), 360.

16. Hunter, *Steamboats*, 201–206.

17. Donald T. Zimmer, "The Ohio River: Pathway to Settlement," in *Transportation in the Early Nation* (Indianapolis: Indiana Historical Society, 1982), 61–88. Stine quoted from *Mixing the Waters: Environment, Politics, and the Building of the Tennessee–Tombigbee Waterway* (Akron, Ohio: University of Akron Press, 1993), 6, *Washington Post* quoted on "wet elephant," 249.

7. Railways

1. John H. White Jr., *The American Railroad Freight Car* (Baltimore, Md.: Johns Hopkins University Press, 1993), 70, 608–611. John S. Blank, *Modern Towing* (Centreville, Md.: Cornell Maritime Press, 1989), 214, 264–272. Don Philips, "Black Diamonds are a Railroad's Best Friend," *Trains*, January 2001, 46–57.

2. On equine capabilities, see John Lienhard, *The Engines of Our Ingenuity* (New York: Oxford University Press, 2000), 93–94. On distances, George Rogers Taylor, *The Transportation Revolution, 1815–1860* (New York: Harper & Row, 1968), 71.

3. Robert C. Post, "Benjamin Dearborn's Railroad Memorial of 1819," *Railroad History* 132 (1975): 84–85. Frederick C. Gamst, "The Context and Significance of America's First Railroad, on Boston's Beacon Hill," *Technology and Culture* 33 (1992): 66–100. Eugene S. Ferguson, "Steam Transportation," in *Technology in Western Civilization*, ed. Melvin Kranzberg and Carroll W. Pursell (New York: Oxford University Press, 1967), 1:284–302.

4. Maurice W. Kirby, *The Origins of Railway Enterprise: The Stockton and Darlington Railway, 1821–1863* (Cambridge, U.K.: Cambridge University Press, 2002). On early British railways, see Michael Freeman, *Railways and the Victorian Imagination* (New Haven, Conn.: Yale University Press, 1999).

5. The classic analysis of urban elites confronting critical choices about technology—which door?—is Julius Rubin, *Canal or Railroad? Imitation and Innovation in the Response to the Erie Canal in Philadelphia, Baltimore, and Boston* (Philadelphia: American Philosophical Society, 1961). On the situation in Charleston, see James E. Vance Jr., *The North American Railroad: Its Origin, Evolution, and Geography* (Baltimore: Johns Hopkins University Press, 1995), 92–97.

6. The B&O has been well served by historians: James D. Dilts, *The Great Road: The Building of the Baltimore & Ohio, the Nation's First Railroad, 1828–1853* (Stanford, Calif.: Stanford University Press, 1993); John F. Stover, *History of the Baltimore and Ohio Railroad* (West Lafayette, Ind.: Purdue University Press, 1987); Herbert H. Harwood, Jr., *Impossible Challenge: The Baltimore and Ohio Railroad in Maryland* (Baltimore, Md.: Barnard, Roberts & Co., 1979). The inception of the South Carolina Canal and Railroad is sketched in Vance, *The North American Railroad*, 92–97.

7. Stephen Salsbury, *The State, the Investor, and the Railroad: The Boston and Albany, 1825–1967* (Cambridge, Mass.: Harvard University Press, 1967). On the Camden & Amboy in brief, see John H. White Jr., *The John Bull: 150 Years a Locomotive* (Washington, D.C.: Smithsonian Institution Press, 1981), 83–103.

8. Roger Barton, "The Camden and Amboy Railroad Monopoly," *Proceedings of the New Jersey Historical Society* (October 1927): 405–418; White, *The John Bull*, 22–34. Smithsonian historian White steamed *The John Bull* along tracks beside the Chesapeake & Ohio Canal on its 150th anniversary in 1981.

9. John F. Stover, *The Routledge Historical Atlas of American Railroads* (New York: Routledge, 1999), 12–21, 68–69.

10. Stover, *The Routledge Historical Atlas*, 16–17. "Climate where hope" quote is from Steven W. Usselman, *Regulating Railroad Innovation: Business, Technology, and Politics in America, 1840–1920* (Cambridge, U.K.: Cambridge University Press, 2002), 8.

11. The mirror-image thesis is set forth in Vance, *The North American Railroad*.

12. John H. White Jr., "Introduction of the Locomotive Safety Truck," *United States National Museum Bulletin* 228 (1961): 118–131. F. Daniel Larkin, *John B. Jervis: An American Engineering Pioneer* (Ames: Iowa State University Press, 1990). On the comparatively obscure Campbell, see John H. White Jr., *A History of the American Locomotive: Its Development, 1830–1880* (New York: Dover Publications, 1979), 450–451. Charles H. Clark, "The Development of the Semiautomatic Freight–Car Coupler, 1863–1893" *Technology and Culture* 13 (1972): 170–208. On the inception of air brakes, Usselman, *Regulating Railroad Innovation*, 276–292.

13. On master mechanics, see Usselman, *Regulating Railroad Innovation*, 70–75. White, *A History of the American Locomotive*, 485–487. John H. White Jr., "Once the Greatest of Builders: The Norris Locomotive Works," *Railroad History* 150 (Spring 1984): 17–56, quote about Long on 18. John K. Brown, *The Baldwin Locomotive Works, 1831–1915* (Baltimore, Md.: Johns Hopkins University Press, 1995).

14. John F. Stover, "One Gauge: How Hundreds of Incompatible Railroads Became a National System," *Invention and Technology* (winter 1993): 55–61; Stover, *The Routledge Historical Atlas*, 26–27, "teamsters, porters, and tavern keepers" quote on 26; Carl H. Scheele, *A Short History of the Mail Service* (Washington, D.C.: Smithsonian Institution Press, 1970), 93, notes that there were eleven gauges in the north alone. Linwood W. Moody, *The Maine Two-Footers* (Berkeley, Calif.: Howell–North Books, 1959); George W. Hilton, *American Narrow Gauge Railroads* (Stanford, Calif.: Stanford University Press, 1990). As with segments of certain old canals, two portions of three-foot-gauge railway in Colorado survive as tourist attractions.

15. John H. White Jr., *The American Railroad Freight Car* (Baltimore, Md.: Johns Hopkins University Press, 1993), 8–35, on traffic and tonnage; White, *The American Railroad Passenger Car* (Baltimore, Md.: Johns Hopkins University Press, 1978).

16. Stover, *The Routledge Historical Atlas*, 20–25. For comparative mileage and investment, see Colleen Dunlavy, *Politics and Industrialization: Early Railroads in the United States and Prussia* (Princeton, N.J.: Princeton University Press, 1994), 28–31. Thomas J. Misa, *A Nation of Steel: The Making of Modern America, 1865–1925* (Baltimore, Md.: Johns Hopkins University Press, 1995).

17. Robert W. Fogel, *Railroads and American Economic Growth: Essays in Econometric History* (Baltimore, Md.: Johns Hopkins University Press, 1964), 235.

18. A rich literature on railways and social psychology includes Leo Marx, *The Machine in the Garden: Technology and the Pastoral Ideal in America* (New York: Oxford University Press, 1964); Wolfgang Schivelbusch, *The Railway Journey: Trains and Travel in the 19th Century* (New York: Urizen Books,

1979); John R. Stilgoe, *Metropolitan Corridor: Railroads and the American Scene* (New Haven, Conn.: Yale University Press, 1983); James A. Ward, *Railroads and the Character of America, 1820–1887* (Knoxville: University of Tennessee Press, 1986), and Sarah H. Gordon, *Passage to Union: How the Railroads Transformed American Life, 1829–1929* (Chicago: Ivan R. Dee, 1996). Nye is quoted from *Electrifying America: Social Meanings of a New Technology* (Cambridge, Mass.: MIT Press, 1990), 85; Staudenmaier from "'The Frailties and Beauties of Technological Creativity': An Interview with John M. Staudenmaier by Robert C. Post," *Invention and Technology* (spring 1993): 24. Steven Usselman, a very fine historian, sees Fogel's view as congruent with his own sense "that railroads emerged not simply as an economically logical choice, but through a contentious and highly emotional political process" (*Regulating Railroad Innovation*, 19–20, n. 10).

19. George Rogers Taylor, *The Transportation Revolution, 1815–1860* (New York: Harper and Row, 1968), esp. 132–175. Albert Fishlow, *American Railroads and the Transformation of the Ante-Bellum Economy* (Cambridge, Mass.: Harvard University Press, 1965). John F. Stover, *Iron Road to the West: American Railroads in the 1850s* (New York: Columbia University Press, 1978). William Cronon, *Nature's Metropolis: Chicago and the Great West* (New York: Norton, 1991).

20. Usselman, *Regulating Railroad Innovation*, 32–51.

21. On railways and the Civil War in brief, see Stover, *The Routledge Historical Atlas*, 28–31.

22. Perhaps the best of many accounts of the Pacific Railroad is David Haward Bain, *Empire Express: Building the First Transcontinental Railroad* (New York: Viking Press, 1999), but for flavor see also Oscar Lewis, *The Big Four: The Story of Huntington, Stanford, Hopkins, and Crocker and the Building of the Central Pacific* (New York: Alfred A. Knopf, 1938), and Maury Klein, *Union Pacific: Birth of a Railroad, 1862–1893* (Garden City, N.Y.: Doubleday, 1987).

23. Carlos Arnaldo Schwantes, *Long Day's Journey: The Steamboat and Stagecoach Era in the Northern West* (Seattle: University of Washington Press, 1999), 273.

24. Stover, *The Routledge Historical Atlas*, 36–49. Ian Bartky, *Selling the True Time: Nineteenth-Century Timekeeping in America* (Stanford, Calif.: Stanford University Press, 2000).

25. Stover, *The Routledge Historical Atlas*, 48–59. William D. Middleton, *When the Steam Railroads Electrified*, 2d. ed. (Bloomington: Indiana University Press, 2001); Gordon D. Friedlander, "Railroad Electrification: Past, Present, Future. History of Systems in the United States," *IEEE Spectrum*, July 1968, 50–89. Kurt C. Schlichting, *Grand Central Terminal: Railroads, Engineering, and Architecture in New York City* (Baltimore, Md.: Johns Hopkins University Press, 2001). Margaret Coel, "A Silver Streak," *Invention and Technology* (fall 1986):

10–17. For a case study of a railway that was bankrupted, see H. Roger Grant, *The North Western: A History of the Chicago and North Western Railway System* (DeKalb: Northern Illinois University Press, 1996).

26. Ernest W. Williams, "Rail and Water Transport," in *Technology in Western Civilization*, ed. Kranzberg and Pursell, 2:137–153. On the switch to diesels in brief, see Maury Klein, "The Diesel Revolution," *Invention and Technology* (winter 1991): 16–22; for detail, Albert Churella, *From Steam to Diesel: Managerial Customs and Organizational Capabilities in the Twentieth-Century Locomotive Industry* (Princeton, N.J.: Princeton University Press, 1998). On the final development of reciprocating steam, along with remarkable illustrations, David Weitzman, *Superpower: The Making of a Steam Locomotive* (Boston: David R. Godine, 1987).

27. Statistics from Bruce E. Seely, "Railroads, Good Roads, and Motor Vehicles: Managing Technological Change," *Railroad History* 155 (1986): 34–63. Stephen Salsbury, *No Way to Run a Railroad: The Untold Story of the Penn Central Crisis* (New York: McGraw Hill, 1982).

28. Seely, "Railroads, Good Roads, and Motor Vehicles."

8. Highways

1. Ben J. Wattenberg, *The Statistical History of the United States. From Colonial Times to the Present* (New York: Basic Books, 1976), 729–730. Bruce E. Seely, "Railroads, Good Roads, and Motor Vehicles: Managing Technological Change," *Railroad History* 155 (1986): 34–63.

2. William L. Richter, *Transportation in America* (Santa Barbara, Calif.: ABC–Clio, 1995), 212–213, 274–276. Seely, "Railroads, Good Roads, and Motor Vehicles." David A. Hounshell, "The Bicycle and Technology in Late Nineteenth Century America," in *Transport Technology and Social Change* ed. Per Sorbom (Stockholm: Teknista Museet, 1980), 175–185.

3. Seely, "Railroads, Good Roads, and Motor Vehicles."

4. Lord Earl Curzon, quoted from *Motor*, January 1919, in James J. Flink, *The Car Culture* (Cambridge, Mass.: MIT Press, 1975), 93.

5. Historians also found it difficult to get out of a Robber Baron mind-set. The leading revisionist was Albro Martin, whose rhetorical style can be sampled in "Light at the End of a Very Long Tunnel: The Railroads and the Historians," *Railroad History* 155 (1986): 15–33. Usselman quoted from *Regulating Railroad Innovation: Business, Technology, and Politics in America, 1840–1920* (Cambridge, U.K.: Cambridge University Press, 2002), 12. On thermal efficiencies, see Lynwood Bryant, "The Beginnings of the Internal Combustion Engine," in *Technology in Western Civilization*, ed. Melvin Kranzberg and Carroll W. Pursell (New York: Oxford University Press, 1967), 1:648–664.

6. John H. White Jr., "The Magic Box: Genesis of the Container," *Railroad History* 158 (spring 1988): 72–93; Volti quoted in *The Facts on File Encyclopedia of Science, Technology, and Society* ed. Volti (New York: Facts on File, 1999), 1058. Jim Giblin, "Are Truckers Lucky? Or Just Better?" *Trains*, November 2001, 54–57.

7. Margaret Creighton points out that the woman most prominent on oceangoing ships was the wooden figurehead, "immobilized from the waist down" ("Sailing Between a Rock and a Hard Place: Navigating Manhood in the 1800s," in Benjamin W. Labaree et al., *America and the Sea: A Maritime History* [Mystic, Conn.: Mystic Seaport, 1998], 293–296). Linda Nieman, *Boomer: Railroad Memoirs* (Berkeley: University of California Press, 1990); Linda Nieman and Lina Bertucci, *Railroad Voices* (Stanford, Calif.: Stanford University Press, 1998); Virginia Scharff, *Taking the Wheel: Women and the Coming of the Motor Age* (New York: The Free Press, 1991), 13. On how fluidity permeated a certain "masculine manger," see Robert C. Post, "Drag Racing," in *Boyhood in America: An Encyclopedia*, ed. Priscilla Ferguson Clement and Jacqueline Reiner (Santa Barbara, Calif.: ABC–Clio, 2001), 1:217–221. On male historians of technology perpetuating male hegemony, Joseph J. Corn, "Tools, Technologies, and Contexts: Interpreting the History of American Technics," in *History Museums in the United States: A Critical Assessment* ed. Warren Leon and Roy Rosenzweig (Urbana: University of Illinois Press, 1989), 237–261.

8. *The Education of Henry Adams*, ed. Ernest Samuels (Boston: Houghton Mifflin, 1974), 330.

9. John B. Rae, *The American Automobile* (Chicago: University of Chicago Press, 1965), 1.

10. Otto Mayr and Robert C. Post, eds., *Yankee Enterprise: The Rise of the American System of Manufactures* rev. ed. (Washington, D.C.: Smithsonian Institution Press, 1995); David A. Hounshell, *From the American System to Mass Production, 1800–1932: The Development of Manufacturing Technology in the U.S.* (Baltimore, Md.: Johns Hopkins University Press, 1984).

11. Ford quoted in Rae, *The American Automobile*, 59.

12. Rae, *The American Automobile*, 59–62. A signal victory for Ford was overturning an 1878 "road engine" patent held by George Selden, a "man before his time" somewhat reminiscent of John Fitch. See Rae, *The American Automobile*, 33–40.

13. James J. Flink, "Unplanned Obsolescence," *Invention and Technology*, Summer 1996, 58–62. J. M. Fenster, "How General Motors Beat Ford," *Audacity* (fall 1992): 50–60.

14. On interurbans in brief, see George W. Hilton, "The Wrong Track," *Invention and Technology* (spring 1993): 47–54; for details, Hilton and John F. Due, *The Electric Interurban Railways in America* (Stanford, Calif.: Stanford University Press, 1960).

15. Bruce E. Seely, *Building the American Highway System: Engineers as Policy Makers* (Philadelphia: Temple University Press, 1987); Mark Rose, *Interstate: Express Highway Politics, 1941–1956* (Lawrence: Regents Press of Kansas, 1979). On the interstate system, Tom Lewis, *Divided Highways: Building the Interstate Highways, Transforming American Life* (New York: Viking, 1997), or, in brief, T. A. Heppenheimer, "The Rise of the Interstates," *Invention and Technology* (fall 1991): 8–14.

16. Donald Davis and Barbara Lorenzkowski, "A Platform for Gender Tensions: Working Women and Riding on Canadian Urban Mass Transit in the 1940s," *Canadian Historical Review* 79 (September 1998): 431–465. Clay McShane, "Gender Wars," chap. 8 in *Down the Asphalt Path: The Automobile and the American City* (New York: Columbia University Press, 1994). On the construction of technology in order to attain specific social aims, see Langdon Winner, "Do Artifacts Have Politics?" reprinted from *Daedelus*, in *The Social Shaping of Technology* ed. Donald MacKenzie and Judy Wajcman (Buckingham, U.K.: Open University Press, 2d ed. 1999), 28–40.

17. Wattenberg, *The Statistical History*, 729, 769. Volti quoted from "A Century of Automobility," *Technology and Culture* 37 (1996): 663–685, on 679.

9. AIRWAYS

1. For differing takes on this question, cf. Rudi Volti, "Why Internal Combustion?" *Invention & Technology* (spring 1990): 42–47; Mickael Haard, ed., *Automobile Engineering in a Dead End: Mainstream and Alternative Developments in the 20th Century* (Goethenberg: Goethenberg University, 1992); Clay McShane, *Down the Asphalt Path: The Automobile and the American City* (New York: Columbia University Press, 1994); Michael B. Schiffer, *Taking Charge: The Electric Automobile in America* (Washington, D.C.: Smithsonian Institution Press, 1994); and David A. Kirsch, *The Electric Vehicle and the Burden of History* (New Brunswick, N.J.: Rutgers University Press, 2000).

2. As Walter Vincenti remarks ("The Technical Shaping of Technology: Real-World Constraints and Technical Logic in Edison's Electric Light System," *Social Studies of Science* 25 [1995]: 565), perpetual motion poses an ultimate challenge to the concept of "sociological explanation." Richard K. Smith shows how everything else about airplane design is contingent on controlling the weight in "The Intercontinental Airliner and the Essence of Airplane Performance, 1929–1939," *Technology and Culture* 24 (1983): 428–449. It should be noted, however, that real-world constraints can be mistaken. In the early 1950s, for example, enthusiasts for the new sport of drag racing were told that the "laws of physics" rendered it impossible to design a "dragster" capable of accelerating to a speed faster than 166 miles-per-hour in the space of 1,320 feet; they remained skeptical, and fifty years later such vehicles can clock more than double that alleged maximum. Robert C. Post, *High Performance: The Culture and Technology of Drag Racing, 1950–2000*, rev. ed. (Baltimore, Md.: Johns Hopkins University Press,. 2001).

3. The most satisfactory survey is Roger E. Bilstein, *Flight in America: From the Wrights to the Astronauts* (Baltimore, Md.: Johns Hopkins University Press, 3d ed. 2001), which is the source for the 8,000 figure (6). In "Eilmer of Malmsbury, An Eleventh Century Aviator: A Case Study in Technological Innovation, Its Context and Tradition," *Technology and Culture* 2 (1961): 97–111, Lynn White marshals evidence that a Benedictine monk "flew with a rigid wing of considerable size" for a distance of some 600 feet sometime between 1000 and 1010 C.E.

4. For the saga of the Wright brothers in brief, see Tom D. Crouch, "How the Bicycle Took Wing," *Invention and Technology* (summer 1986): 10–16, and "Kill Devil Hills, 17 December 1903," *Technology and Culture* 40 (1999): 594–598; for detail, Crouch, *The Bishop's Boys: A Life of Wilbur and Orville Wright* (New York: Norton, 1989); and for their milieu, Crouch, *A Dream of Wings: Americans and the Airplane, 1875–1905* (New York: Norton, 1981).

5. "Government agent" quote from Noriss C. Hetherington, "The Langley and Wright Aero Accidents: Two Responses to Early Aeronautical Innovation and Government Patronage," in *Innovation and the Development of Flight* ed. Roger Launius (College Station: Texas A&M University Press, 1999), 18–51, on 22. On Langley's misadventures, a rich account appears in Mark Sullivan's classic narrative from the 1920s, *Our Times 1900–1925*, vol. 2, *America Finding Herself* (New York: Charles Scribner's Sons, 1971), 557–568; only the Smithsonian continued to assert Langley's priority, one of its several embarrassments involving aeronautics, the most recent involving the contested storyline for an exhibit of the *Enola Gay*, the B–29 that loosed the atomic bomb on Hiroshima in 1945.

6. Joseph J. Corn, *The Winged Gospel: America's Romance with Aviation, 1900–1950* (New York: Oxford University Press, 1983). On "aeronauts," see Corn, 71–90; Bilstein, *Flight in America*, 17–31; and Julie Wosk, "Women and Aviation," chap. 6 in *Woman and the Machine: Representations from the Spinning Wheel to the Electronic Age* (Baltimore, Md.: Johns Hopkins University Press, 2001).

7. Corn quote from *The Winged Gospel*, 12. Bilstein, *Flight in America*, 15–26. John William Ward, "The Meaning of Lindbergh's Flight," *American Quarterly* 10 (1958): 3–16. The irony with Rogers was that he *did* lose his life a year later while stunting.

8. Corn, "'A New Sign in the Heavens': The Prophetic Creed of Flight," chap. 2 in *The Winged Gospel*; Ryan quoted in Corn on 12, Northrop on 66.

9. Peter L. Jakab and Rick Young, eds., *The Published Writings of Wilbur and Orville Wright* (Washington, D.C.: Smithsonian Institution Press, 2000). Timothy Moy, *War Machines: Transforming Technologies in the U.S. Military, 1920–1940* (College Station: Texas A&M University Press, 2001). Halberstam quoted from *War in a Time of Peace: Bush, Clinton, and the Generals* (New York: Simon and Schuster, 2001), on 457.

10. Robert van der Linden, *Airlines and Airmail: The Post Office and the Birth of the Commercial Aviation Industry* (Lexington: University of Kentucky Press, 2002); Janet R. Daly Bednarek, *America's Airports: Airfield Development, 1918–1947* (College Station: Texas A&M University Press, 2001).

11. Ben J. Wattenberg, *The Statistical History of the United States from Colonial Times to the Present* (New York: Basic Books, 1976), 769. On the inception of commercial aviation in brief, see Bilstein, *Flight in America*, 47–75; for detail, R.E.G. Davies, *Airlines of the United States since 1914* (New York: Rowman, 1972). But lest one forget where the heart of the technological story now lies, in the "military dimension," Anne Millbrooke's compendious *Aviation History* (Englewood, Colo.: Jepperson Sanderson, 1999), devotes relatively little space to what is called "peacetime aviation."

12. John B. Rae, *Climb to Greatness: The American Aircraft Industry, 1920–1960* (Cambridge, Mass.: MIT Press, 1968). Leary quoted from "The Search for an Instrument Landing System, 1918–48," in *Innovation and the Development of Flight*, ed. Launius, 80–99, on 97; McFarland quoted from "Higher, Faster, and Farther: Fueling the Aeronautical Revolution, 1919–45," in ibid., 100–131, on 100.

13. Examples of these various approaches include Edward W. Constant, II, *The Origins of the Turbojet Revolution* (Baltimore, Md.: Johns Hopkins University Press, 1980); Walter G. Vincenti, *What Engineers Know and How they Know It: Analytical Studies from Aeronautical History* (Baltimore, Md.: Johns Hopkins University Press, 1990); Eric Shatzberg, *Wings of Wood, Wings of Metal: Culture and Technical Choice in American Airplane Materials, 1914–1945* (Princeton, N.J.: Princeton University Press, 1999); and Corn, *The Winged Gospel*, as well as two older works, William Fielding Ogburn, *The Social Effects of Aviation* (Boston: Houghton Mifflin, 1946); and Irving B. Holley, *Ideas and Weapons: Exploitation of the Aerial Weapon by the United States during World War I* (New Haven, Conn.: Yale University Press, 1953). In "Aviation History in Wider View," *Technology and Culture* 30 (1989): 643–656, James Hanson notes that this specialty is "regarded by many scholars, both insiders and outsiders, as a field too full of 'enthusiasm'" (647), but Bilstein is not at in the least defensive, declaring that "we are legion, and the Force is with us" ("Aerospace Historians, Aerospace Enthusiasts," *Technology and Culture* 28 [1987]: 125).

14. Deborah G. Douglas, "Three-Miles-a-Minute: The National Advisory Committee for Aeronautics and the Development of the Modern Airliner," in *Innovation and the Development of Flight*, ed. Launius, 154–165. Roland quoted from *Model Research: The National Advisory Committee for Aeronautics 1915–1958*, 2 vols. (Washington, D.C.: National Aeronautics and Space Administration, 1985), xiii. This is the indispensable history of NACA, and yet Roland is exceedingly critical, particularly in depicting its "public posturing" in the face of political insecurity in much the same terms as do critics of the Army Corps of Engineers.

15. Allan Sloan, "Planes, Trains and Politicians," *Newsweek*, 2 October, 2002, 49.

CONCLUSION

1. Martin Wachs quoted from "The Evolution of Transportation Policy in Los Angeles: Images of Past Policies and Future Prospects," in Allen J. Scott and Edward W. Soja, eds., *The City: Los Angeles and Urban Theory and the End of the Twentieth Century* (Berkeley: University of California Press, 1996), 106–159, on 107.

2. Robert C. Post, "The Page Locomotive: Federal Sponsorship of Invention in Mid-19th-Century America," *Technology and Culture* 13 (1972): 140–169, and Post, "To Harness the Forces of Nature," chap. 5 in *Physics,Patents, and Politics: A Biography of Charles Grafton Page* (New York: Science History Publications USA, 1976).

3. Roth quoted from "Mulholland Highway and the Engineering Culture of Los Angeles in the 1920s," *Technology and Culture* 40 (1999): 545–575, on 531. The "democratic impulse" argument is set forth in Scott L. Bottles, *Los Angeles and the Automobile: The Making of the Modern City* (Berkeley: University of California Press, 1987). Accounts of the alleged conspiracy have been published in dozens of books, the most shrill often by the most reputable presses, for example: Jane Holtz Kay, *Asphalt Nation: How the Automobile Took Over America and How We Can Take It Back* (Berkeley: University of California Press, 1997), and Stephen B. Goddard, *Getting There: The Epic Struggle between Road and Rail in the American Century* (Chicago: University of Chicago Press, 1994). For the historiography in extended dimension, see Robert C. Post, "Images of the Pacific Electric: Why Memories Matter," *Railroad History* 179 (1998): 31–68.

4. Robert C. Post, "The Myth Behind the Streetcar Revival," *American Heritage*, May 1998, 95–100.

5. Mike Davis, "L.A. Transit Apartheid: Runaway Train Crushes Buses," *The Nation*, 18 September, 1995, 270– 274, quote on 273.

6. This is according to an article in the American magazine *Railfan & Railroad* (Ron Ziel, "Big Steam's Last Stand," August 2002, 44–51). The eminent British historian, Lord Asa Briggs, writes that "it is a feature of technologies that they captivate most when they are disappearing or dead" (*The Power of Steam* [Chicago: University of Chicago Press, 1982], 12). It is worth noting that every sort of conveyance has a cadre of "fans," men (and they are overwhelmingly male) who carry on a love affair with locomotives, trolley cars, towboats, ore-boats, automobiles, airplanes, buses, bicycles, what have you. Some of them believe that it was sinful if not indeed criminal for railroads to have put canals out of business, or for diesel locomotives to have replaced steam locomotives.

Suggestions for Further Reading

Note: This list does not recapitulate every reference in the endnotes, which sometimes do not directly address travel, transport, and technology. Likewise omitted are references to periodicals. A comprehensive bibliography would run to thousands of titles, I daresay tens of thousands. Because this is a book*let*, the list cuts off with a nice round number, 200 references, mainly recent imprints.

Reference Books

Albion, Robert G, *Naval & Maritime History: An Annotated Bibliography*. 4th ed. Mystic., Conn.: Marine Historical Assn., 1972.

Berger, Michael L., *The Automobile in American History: A Reference Guide*. Westport, Conn.: Greenwood Press, 2001.

Bryant, Keith L., Jr., ed. *Encyclopedia of American Business History and Biography: Railroads in the Age of Regulation, 1900–1980*. New York: Facts on File, 1988.

Chapelle, Howard I. *The National Watercraft Collection*. 2nd ed. Camden, Maine: International Marine Publishing Co., 1976.

Frey, Robert L., ed. *Encyclopedia of American Business History and Biography: Railroads in the Nineteenth Century*. New York: Facts on File, 1988.

Hindle, Brooke. *Technology in Early America: Needs and Opportunities for Study*. Chapel Hill: University of North Carolina Press, 1966.

Kemp, Peter, ed. *The Oxford Companion to Ships and the Sea*. New York: Oxford University Press, 1976.

McShane, Clay. *The Automobile: A Chronology of Its Antecedents, Development, and Impact*. Westport, Conn.: Greenwood Press, 1995.

Millbrooke, Anne. *Aviation History*. Englewood, Colo.: Jepperson Sanderson, 1999.

Richter, William L. *Transportation in America*. Santa Barbara, Calif.: ABC–Clio, 1995.

Stover, John F. *The Routledge Historical Atlas of American Railroads*. New York: Routledge, 1999.

Trinder, Barrie. *The Blackwell Encyclopedia of Industrial Archaeology*. Cambridge, Mass.: Basil Blackwell Ltd., 1992.

Volti, Rudi, ed. *The Facts on File Encyclopedia of Science, Technology, and Society*. New York: Facts on File, 1999.

Wattenberg, Ben J. *The Statistical History of the United States From Colonial Times to the Present*. New York: Basic Books, 1976.

SURVEYS, MONOGRAPHS, ANTHOLOGIES

Albion, Robert G. *The Rise of New York Port (1815–1860)*. New York: Charles Scribner's Sons, 1939.

Albion, Robert G. *Square-Riggers on Schedule: The New York Sailing Packets to England, France, and the Cotton Ports*. Princeton, N.J.: Princeton University Press, 1938.

Ambler, Charles Henry. *A History of Transportation in the Ohio Valley*. Glendale, Calif.: Arthur H. Clark Co., 1932.

Baer, Christopher T. *Canals and Railroads of the Mid-Atlantic States, 1800–1860*. Wilmington, Del.: Regional Economic Research Center, Eleutherian Mills-Hagley Foundation, 1981.

Bain, David Haward. *Empire Express: Building the First Transcontinental Railroad*. New York: Viking Press, 1999.

Baldwin, Leland D. *The Keelboat Age on Western Waters*. Pittsburgh: University of Pittsburgh Press, 1941.

Barry, James P. *Ships of the Great Lakes: 300 Years of Navigation*. (Berkeley, Calif.: University of California Press, 1973.

Bartky, Ian R. *Selling the True Time: Nineteenth-Century Timekeeping in America*. Stanford, Calif.: Stanford University Press, 2000.

Basalla, George. *The Evolution of Technology*. Cambridge, U.K.: Cambridge University Press, 1988).

Bednarek, Janet R. Daly. *America's Airports: Airfield Development, 1918–1947*. College Station: Texas A&M University Press, 2001.

Belasco, Warren. *Americans on the Road: From Autocamp to Motel, 1910–1945*. Cambridge, Mass.: MIT Press, 1979.

Berger, Michael L. *The Devil Wagon in God's Country: The Automobile and Social Change in Rural America, 1893–1929*. Hamden, Conn.: Archon Books, 1979.

Berk, Gerald. *Alternative Tracks: The Constitution of the American Industrial Order, 1865–1917*. Baltimore, Md.: Johns Hopkins University Press, 1994.

Bilstein, Roger E. *The Enterprise of Flight: The American Aviation and Aerospace Industry*. Washington, D.C.: Smithsonian Institution Press, 2001.

Bilstein, Roger E. *Flight in America: From the Wrights to the Astronauts*. 3rd ed. Baltimore, Md.: Johns Hopkins University Press, 2001.

Blank, John S. *Modern Towing*. Centreville, Md.: Cornell Maritime Press, 1989.

Boorstin, Daniel. *The Americans: The National Experience*. New York: Vintage, 1965.

Bottles, Scott L. *Los Angles and the Automobile: The Making of the Modern City*. Berkeley, Calif.: University of California Press, 1987.

Brown, Alexander Crosby. *Juniper Waterway: A History of the Albemarle and Chesapeake Canal*. Charlottesville: University Press of Virginia, 1981.

Brown, John K. *The Baldwin Locomotive Works, 1831–1915*. Baltimore, Md.: Johns Hopkins University Press, 1995.

Bulliet, Richard W. *The Camel and the Wheel*. Cambridge, Mass.: Harvard University Press, 1975.

Calhoun, Daniel. *The American Civil Engineer: Origins and Conflict*. Cambridge, Mass.: MIT Press, 1960.

Cecelski, David S. *The Waterman's Song: Slavery and Freedom in Maritime North Carolina*. Chapel Hill: University of North Carolina Press, 2001.

Chandler, Alfred D., Jr. *Giant Enterprise: Ford, General Motors, and the Automobile Industry*. New York: Harcourt, Brace and World, 1964.

Chandler, Alfred D., Jr. *The Visible Hand: The Managerial Revolution in American Business*. Cambridge, Mass.: Harvard University Press 1977.

Chapelle, Howard I. *The Baltimore Clipper: Its Origins and Development*. Salem, Mass.: Marine Research Society, 1930.

Churella, Albert J. *From Steam to Diesel: Managerial Customs and Organizational Capabilities in the Twentieth-Century Locomotive Industry.* Princeton, N.J.: Princeton University Press, 1998.

Clark, Mary Stetson. *The Old Middlesex Canal.* Melrose, Mass.: Hilltop Press, 1974.

Condit, Carl W. *The Pioneer Stage of Railroad Electrification.* Philadelphia: American Philosophical Society, 1977.

Constant, Edward W., II. *The Origins of the Turbojet Revolution.* Baltimore, Md.: Johns Hopkins University Press, 1980.

Corn, Joseph J. *The Winged Gospel: America's Romance with Aviation, 1900 1950.* New York: Oxford University Press, 1983.

Cowan, Ruth Schwartz. *A Social History of American Technology.* New York: Oxford University Press, 1997.

Cressy, David. *Coming Over: Migration and Communication Between England and New England in the Seventeenth Century.* Cambridge: Cambridge University Press, 1987.

Cronon, William. *Nature's Metropolis: Chicago and the Great West.* New York: Norton, 1991.

Crouch, Tom D. *The Bishop's Boys: A Life of Wilbur and Orville Wright.* New York: Norton, 1989.

Crouch, Tom D. *A Dream of Wings: Americans and the Airplane.* New York: Norton, 1981.

Cutler, Carl C. *Greyhounds of the Sea: The Story of the American Clipper Ship.* New York: G. P. Putnam's Sons, 1930.

Davies, R.E.G. *Airlines of the United States since 1914.* New York: Rowman, 1972.

Davis, Donald Finley. *Conspicuous Production: Automobiles and Elites in Detroit, 1899–1933.* Philadelphia: Temple University Press, 1988.

Dilts, James D. *The Great Road: The Building of the Baltimore & Ohio, the Nation's First Railroad, 1828–1853.* Stanford: Stanford University Press, 1993.

Drago, Harry Sinclair. *Canal Days in America.* New York: Brimhall House, 1972.

Dunbar, Seymour. *A History of Travel in America*, 4 vols. Indianapolis, Ind.: Bobbs-Merrill, 1915.

Dunlavy, Colleen. *Politics and Industrialization: Early Railroads in the United States and Prussia*. Princeton: Princeton University Press, 1994.

Dunn, James A., Jr. *Miles to Go: European and American Transportation Policies*. Cambridge, Mass.: MIT Press, 1981.

Eastman, Joel W. *Styling vs. Safety: The American Automobile Industry and the Development of Automotive Safety, 1900–1966*. Lanham, Md.: University Press of America, 1984.

Fishlow, Albert. *American Railroads and the Transformation of the Ante-Bellum Economy*. Cambridge, Mass.: Harvard University Press, 1965.

Flexner, James Thomas. *Steamboats Come True: American Inventors in Action*. Boston: Little, Brown and Co., 1978. Orig. pub. 1944 as *Inventors in Action*.

Flink, James J. *America Adopts the Automobile*. Cambridge, Mass.: MIT Press, 1970.

Flink, James J. *The Automobile Age*. Cambridge, Mass.: MIT Press, 1988.

Flink, James J. *The Car Culture*. Cambridge, Mass.: MIT Press, 1975.

Fogel, Robert W. *Railroads and American Economic Growth*. Baltimore: Johns Hopkins University Press, 1964.

Foster, Mark S. *From Streetcar to Superhighway: American City Planners and Urban Transportation, 1900–1940*. Philadelphia: Temple University Press, 1981.

Frederick, J.V. *Ben Holliday, the Stagecoach King: A Chapter in the Development of Transcontinental Transportation*. Lincoln: University of Nebraska Press, 1989.

Freeman, Michael. *Railways and the Victorian Imagination*. New Haven, Conn.: Yale University Press, 1999.

Gamst, Frederick C., ed. David J. Diephouse and John C. Decker, trans. *Early American Railroads: Franz Anton von Gerstner's* Die innern Communicationen *(1842–1843)*. Stanford, Calif.: Stanford University Press, 1997. Trans. of German edition.

Gibbs-Smith, Charles H. *The Invention of the Aeroplane, 1799–1909*. London: Her Majesty's Stationery Office, 1966.

Gilfillan, S.C. *Inventing the Ship*. Chicago: Follett Publishing, 1935.

Goddard, Stephen B. *Getting There: The Epic Struggle Between Road and Rail in the American Century*. Chicago: University of Chicago Press, 1994.

Goldenberg, Joseph. *Shipbuilding in Colonial America*. Charlottesville: University of Virginia Press, 1976.

Goodrich, Carter. *Government Promotion of American Canals and Railroads, 1800–1890.* New York: Columbia University Press, 1960.

Gordon, Sarah H. *Passage to Union: How the Railroads Transformed American Life, 1829–1929.* Chicago: Ivan R. Dee, 1996.

Gorn, Michael H. *Expanding the Envelope: Flight Research at NACA and NASA.* Lexington: University Press of Kentucky, 2001.

Grant, H. Roger. *The North Western: A History of the Chicago and North Western Railway System.* DeKalb: Northern Illinois University Press, 1996.

Gray, Ralph D. *The Great National Waterway: A History of the Chesapeake & Delaware Canal, 1769–1965.* Urbana: University of Illinois Press, 1969.

Greenleaf, William. *Monopoly on Wheels: Henry Ford and the Selden Automobile Patent Suit.* Detroit: Wayne State University Press, 1961.

Haard, Mikael, ed. *Automobile Engineering in a Dead End: Mainstream and Alternative Developments in the 20th Century.* Goethenberg, Sweden: Goethenberg University, 1992.

Haites, Erik F., James Mak, and Gary M. Walton. *Western River Transportation: The Era of Internal Development, 1810–1860.* Baltimore, Md.: Johns Hopkins University Press, 1975.

Harwood, Herbert H., Jr. *Impossible Challenge: The Baltimore and Ohio Railroad in Maryland.* Baltimore, Md.: Barnard, Roberts & Co., 1979.

Headrick, Daniel R. *Tools of Empire: Technology and European Imperialism in the Nineteenth Century.* New York: Oxford University Press, 1981.

Heckman, Richard. ed. *Yankees Under Sail.* Dublin, N.H.: Yankee, Inc., 1968.

Hill, Forrest G. *Roads, Rails, and Waterways: The Army Engineers and Early Transportation.* Norman: University of Oklahoma Press, 1957.

Hilton, George W. *American Narrow Gauge Railroads.* Stanford, Calif.: Stanford University Press, 1990)

Hilton, George W., and John F. Due. *The Electric Interurban Railways in America.* Stanford, Calif.: Stanford University Press, 1960.

Hindle, Brooke. *Emulation and Innovation.* New York: New York University Press, 1981.

Hindle, Brooke, ed. *Material Culture of the Wooden Age.* Tarrytown, N.Y.: Sleepy Hollow Press, 1981.

Holley, Irving Brinton. *Ideas and Weapons: Exploitation of the Aerial Weapon by the United States During World War I.* New Haven: Yale University Press, 1953.

Horwitch, Mel. *Clipped Wings: The American SST Conflict.* Cambridge, Mass.: MIT Press, 1982.

Hounshell, David A. *From the American System to Mass Production, 1800–1932: The Development of Manufacturing Technology in the U.S.* Baltimore, Md: Johns Hopkins University Press, 1984.

Hunter, Louis C. *Steamboats on the Western Rivers: An Economic and Technological History.* Cambridge, Mass.: Harvard University Press, 1949.

Indiana Historical Society. *Transportation in the Early Nation.* Indianapolis, Ind.: Indiana Historical Society, 1982.

Jackson, W. Turrentine. *Wagon Roads West: A Study of Federal Road Surveys and Construction in the Trans-Mississippi West, 1848–1869.* Berkeley: University of California Press, 1965.

Jakab, Peter L., and Rick Young, eds. *The Published Writings of Wilbur and Orville Wright.* Washington, D.C.: Smithsonian Institution Press, 2000.

Johnston, Paul F. *Steam and the Sea.* Salem, Mass.: Peabody Museum of Salem, 1983.

Jordan, Philip D. *The National Road.* Indianapolis: Bobbs Merrill, 1948.

Kay, Jane Holtz. *Asphalt Nation: How the Automobile Took Over America and How We Can Take It Back.* Berkeley: University of California Press, 1997.

Kemp, Emory L. *The Great Kanawha Navigation.* Pittsburgh, Pa.: University of Pittsburgh Press, 2000.

Kirby, Maurice W. *The Origins of Enterprise: The Stockton and Darlington Railway, 1821–1863.* Cambridge: Cambridge University Press, 2002.

Kirsch, David A. *The Electric Vehicle and the Burden of History.* (New Brunswick, N.J.: Rutgers University Press, 2000.

Klein, Maury. *Unfinished Business: The Railroad in American Life.* Hanover, N.H.: University Press of New England, 1994.

Klein, Maury. *Union Pacific: Birth of a Railroad, 1862–1893.* Garden City, N.Y.: Doubleday, 1987.

Kranzberg, Melvin, and Carroll W. Pursell Jr., eds. *Technology in Western Civilization,* 2 vols. New York: Oxford University Press, 1967.

Labaree, Benjamin W., William M. Fowler, Jr., Edward W. Sloan, John B. Hattendorf, Jeffrey J. Safford, and Andrew W. German. *America and the Sea: A Maritime History.* Mystic, Conn.: Mystic Seaport, 1998.

Laing, Alexander. *American Ships.* New York: American Heritage Press, 1971.

Larkin, F. Daniel. *John B. Jervis: An American Engineering Pioneer*. Ames: Iowa State University Press, 1990.

Larson, John L. *Bonds of Enterprise: John Murray Forbes and Western Development in America's Railway Age*. Rev. ed. Iowa City: University of Iowa Press, 2001).

Larson, John L. *Internal Improvement: National Public Works and the Promise of Popular Government in the Early United States*. Chapel Hill: University of North Carolina Press, 2001.

Lass, William E. *A History of Steamboating on the Upper Missouri*. Lincoln: University of Nebraska Press, 1962.

Law, John. *Aircraft Stories: Decentering the Object in Technoscience*. Durham, N.C.: Duke University Press, 2001.

Lesstrang, Jacques. *Seaway: The Untold Story of North America's Fourth Seacoast*. Seattle: Salisbury Press, 1976.

Lewis, David L. and Laurence Goldstein, eds. *The Automobile and American Culture*. Ann Arbor: University of Michigan Press, 1983.

Lewis, Oscar. *The Big Four: The Story of Huntington, Stanford, Hopkins, and Crocker and the Building of the Central Pacific*. New York: Alfred A. Knopf, 1938.

Lewis, Tom. *Divided Highways: Building the Interstate Highways, Transforming American Life*. New York: Viking, 1997.

Lubbock, A. Basil. *The Down Easters: American Deep-Water Sailing Ships, 1869–1929*. 2d ed. Glasgow, Scotland: J. Brown & Son, 1930.

Martin, Albro. *Enterprise Denied: Origins of the Decline of American Railroads, 1897–1917*. New York: Columbia University Press, 1971.

Martin, Albro. *Railroads Triumphant: The Growth, Rejection & Rebirth of a Vital American Force*. New York: Oxford University Press, 1992.

Marx, Leo. *The Machine in the Garden: Technology and the Pastoral Ideal in America*. New York: Oxford University Press, 1964.

Marx, Leo,. and Merritt Roe Smith, eds. *Does Technology Drive History: The Dilemma of Technological Determinism*. Cambridge, Mass.: MIT Press, 1995.

McCullough, David. *The Great Bridge*. New York: Avon Books, 1972.

McCullough, Robert, and Walter Leuba. *The Pennsylvania Main Line Canal*. York, Pa.: American Canal and Transportation Center, 1973.

Middleton, William D. *When the Steam Railroads Electrified*. 2d. ed. Bloomington, Ind.: Indiana University Press, 2001.

Miller, Ronald, and David Sawers. *The Technological Development of Modern Aviation*. New York: Praeger, 1970.

Misa, Thomas J. *A Nation of Steel: The Making of Modern America, 1865–1925*. Baltimore, Md.: Johns Hopkins University Press, 1995.

Moody, Linwood W. *The Maine Two-Footers*. Berkeley, Calif.: Howell-North, 1959.

Morison, Elting. *From Know-How to Nowhere: The Development of American Technology*. New York: Basic Books, 1974.

Morison, Samuel Eliot. *The Maritime History of Massachusetts, 1783–1860*. Boston: Houghton Mifflin, 1921.

Moy, Timothy. *War Machines: Transforming Technologies in the U.S. Military, 1920–1940*. College Station: Texas A&M University Press, 2001.

Nader, Ralph. *Unsafe at Any Speed: The Designed-In Dangers of the American Automobile*. New York: Grossman, 1965.

Nieman, Linda. *Boomer: Railroad Memoirs*. Berkeley, Calif.: University of California Press, 1990.

Nieman, Linda, and Lina Bertucci. *Railroad Voices*. Stanford, Calif.: Stanford University Press, 1998.

North, Douglass C. *The Economic Growth of the United States, 1790–1860*. New York: Norton, 1966.

Nye, David E. *Electrifying America: Social Meanings of a New Technology*. Cambridge, Mass.: MIT Press, 1990.

Ogburn, William F. *The Social Effects of Aviation*. Boston, Mass.: Houghton Mifflin, 1946.

Oldenziel, Ruth. *Making Technology Masculine: Men, Women, and Modern Machines in America, 1870–1945*. Amsterdam: Amsterdam University Press, 2000.

Olson, Sherry H. *The Depletion Myth: A History of Railroad Use of Timber*. Cambridge, Mass.: Harvard University Press, 1971.

Post, Robert C. *Street Railways and the Growth of Los Angeles*. San Marino, Calif.: Golden West Books, 1989.

Post, Robert C. *The Tancook Whalers: Origins, Rediscovery, Revival*. Bath, Maine: Maine Maritime Museum, 1986.

Prager, Frank D., ed. *The Autobiography of John Fitch*. Philadelphia: American Philosophical Society, 1976.

Pursell, Carroll. *The Machine in America: A Social History of Technology*. Baltimore, Md.: Johns Hopkins University Press, 1995.

Rae, John B. *The American Automobile: A Brief History*. Chicago: University of Chicago Press, 1965.

Rae, John B. *Climb to Greatness: The American Aircraft Industry, 1920–1960*. Cambridge, Mass.: MIT Press, 1968.

Rae, John B. *The Road and the Car in American Life*. Cambridge, Mass.: MIT Press, 1971.

Rose, Mark. *Interstate: Express Highway Politics, 1941–1956*. Lawrence: Regents Press of Kansas, 1979.

Rubin, Julius. *Canal or Railroad? Imitation and Innovation in the Response to the Erie Canal in Philadelphia, Baltimore, and Boston*. Philadelphia: American Philosophical Society, 1961.

Sale, Kirkpatrick.,*The Fire of Genius: Robert Fulton and the American Dream*. New York: Free Press, 2001.

Salsbury, Stephen. *No Way to Run a Railroad: The Untold Story of the Penn Central Crisis*. New York: McGraw Hill, 1982.

Salsbury, Stephen. *The State, the Investor, and the Railroad: The Boston & Albany, 1825–1867*. Cambridge, Mass.: Harvard University Press, 1967.

Sanderlin, Walter S. *The Great National Project: A History of the Chesapeake and Ohio Canal*. Baltimore, Md.: Johns Hopkins University Press, 1946.

Saunders, Richard. *Merging Lines: American Railroads, 1900–1970*. DeKalb: Northern Illinois University Press, 2001.

Scharff, Virginia. *Taking the Wheel: Women and the Coming of the Motor Age*. New York: Free Press, 1991.

Schatzberg, Eric. *Wings of Wood, Wings of Metal: Culture and Technical Choice in American Airplane Materials, 1914–1945*. Princeton: Princeton University Press, 1999.

Scheele, Carl H. *A Short History of the Mail Service*. Washington, D.C.: Smithsonian Institution Press, 1970.

Schieber, Harry N. *Ohio Canal Era: A Case Study of Government and the Economy, 1820–1861*. Athens: Ohio University Press, 1969.

Schiffer, Michael Brain. *Taking Charge: The Electric Automobile in America*. Washington, D.C.: Smithsonian Institution Press, 1994.

Schivelbusch, Wolfgang. *The Railway Journey: Trains and Travel in the 19th Century*. New York: Urizen Books, 1979. Trans. from the German by Anselm Hollo.

Schlichting, Kurt C. *Grand Central Terminal: Railroads, Engineering, and Architecture in New York City*. Baltimore, Md.: Johns Hopkins University Press, 2001.

Schwantes, Carlos Arnaldo. *Long Day's Journey: The Steamboat and Stagecoach Era in the Northern West*. Seattle: University of Washington Press, 1999.

Seely, Bruce E. *Building the American Highway System: Engineers as Policy Makers*. Philadelphia: Temple University Press, 1987.

Shallat, Todd. *Structures in the Stream: Water, Science, and the Rise of the U.S. Army Corps of Engineers*. Austin: University of Texas Press, 1994.

Shaw, Ronald E. *Canals for a Nation: The Canal Era in the United States, 1790–1860*. Lexington: University Press of Kentucky, 1990.

Shaw, Ronald E. *Erie Water West: A History of the Erie Canal, 1792–1854*. Lexington: University of Kentucky Press, 1966.

Sheriff, Carol. *The Artificial River: The Erie Canal and the Paradox of Progress, 1817–1862*. New York: Hill & Wang, 1996.

Silk Gerald, et al. *Automobile and Culture*. New York: Harry Abrams, Inc., 1984.

Smith, Henry Ladd. *The History of Commercial Aviation in the United States*. New York: Knopf, 1942.

Singer, Charles, E. J. Holmyard, A. R. Hall, and Trevor I. Williams, eds. *A History of Technology*, 5 vols. New York: Oxford University Press, 1954–1958.

Sobel, Dava. *Longitude*. London: Fourth Estate, 1996.

Stilgoe, John R. *Metropolitan Corridor: Railroads and the American Scene*. (New Haven: Yale University Press, 1983.

Stine, Jeffrey K. *Mixing the Waters: Environment, Politics, and the Building of the Tennessee–Tombigbee Waterway*. Akron, Ohio: University of Akron Press, 1993.

Stover, John F. *American Railroads*. Chicago: University of Chicago Press, 1961.

Stover, John F. *History of the Baltimore and Ohio Railroad*. West Lafayette, Ind.: Purdue University Press, 1987.

Stover, John F. *Iron Road to the West: American Railroads in the 1850s*. New York: Columbia University Press, 1978.

Stover, John F. *Transportation in American History*. Washington, D.C.: The American Historical Association, 1970.

Struik, Dirk. *Yankee Science in the Making*. Boston: Little, Brown and Co., 1948.

Taylor, George Rogers. *The Transportation Revolution, 1815–1860.* 1951. Reprint, New York: Harper and Row, 1968.

Trinder, Barrie. *The Iron Bridge.* Telford: Ironbridge Gorge Museum Trust, 1979.

Tyler, David Budlong. *Steam Conquers the Atlantic.* New York: D. Appleton-Century Co., 1939.

Usselman, Steven W. *Regulating Railroad Innovation: Business, Technology, and Politics in America, 1840–1920.* Cambridge, U.K.: Cambridge University Press, 2002.

Vance, James E., Jr. *The North American Railroad: Its Origin, Evolution, and Geography.* Baltimore, Md.: Johns Hopkins University Pres, 1995.

Van der Linden, Robert. *Airlines and Airmail: The Post Office and the Birth of the Commercial Aviation Industry.* Lexington: University of Kentucky Press, 2002.

Vincenti, Walter. *What Engineers Know and How they Know It: Analytical Studies from Aeronautical History.* Baltimore, Md.: Johns Hopkins University Press, 1990.

Ward, James A. *J. Edgar Thompson: Master of the Pennsylvania.* Westport, Conn.: Greenwood Press, 1980.

Ward, James A. *Railroads and the Character of America, 1820–1887.* Knoxville: University of Tennessee Press, 1986.

Weitzman, David. *Superpower: The Making of a Steam Locomotive.* Boston, Mass.: David R. Godine, 1987.

Welke, Barbara Young. *Recasting American Liberty: Race, Law, and the Railroad Revolution, 1865–1920.* Cambridge, U.K.: Cambridge University Press, 2001.

White, John H., Jr. *The American Railroad Freight Car.* Baltimore, Md.: Johns Hopkins University Press, 1993.

White, John H., Jr. *The American Railroad Passenger Car.* Baltimore, Md.: Johns Hopkins University Press, 1978.

White, John H., Jr. *A History of the American Locomotive: Its Development, 1830–1880.* New York: Dover Publications, 1979. Orig. pub. by Johns Hopkins University Press in 1968 as *American Locomotives: An Engineering History, 1830–1880.*

White, John H., Jr. *The John Bull: 150 Years a Locomotive.* Washington, D.C.: Smithsonian Institution Press, 1981.

White, John H., Jr. *A Short History of American Locomotive Builders in the Steam Era.* Washington, D.C.: Bass, Inc., 1982.

White, Lynn, Jr. *Medieval Technology and Social Change.* Oxford, U.K.: Oxford University Press, 1962.

Wik, Reynold W. *Henry Ford and Grass Roots America.* Ann Arbor: University of Michigan Press, 1972.

Williams, Trevor I., ed. *A History of Technology: The Twentieth Century.* 2 vols. Oxford: Oxford University Press, 1978.

Winner, Langdon. *Autonomous Technology: Technics-out-of-Control as a Theme in Political Thought.* Cambridge, Mass.: MIT Press, 1977.

Wood, Richard G. *Stephen Harriman Long, 1784–1864: Army Engineer, Explorer, Inventor.* Glendale, Calif.: Arthur H. Clark Co., 1966.

Wosk, Julie. *Woman and the Machine: Representations from the Spinning Wheel to the Electronic Age.* Baltimore, Md.: Johns Hopkins University Press, 2001.

Yates, Brock. *The Decline and Fall of the American Automobile Industry.* New York: Empire Books, 1983.

WAKE TECHNICAL COMMUNITY COLLEGE LIBRARY
9101 FAYETTEVILLE ROAD
RALEIGH, NORTH CAROLINA 27603

WITHDRAWN

DATE DUE

MAR 2 2 '05			
JUN 2 2 '06			
MAR 1 4 2007			
APR 1 9 2007			
MAR 0 9 2013			

JUL '04